SIAN E

ANCIENT AGRICULTURAL IMPLEMENTS

WITHDRAWN

THE
PAUL HAMLYN
LIBRARY

DONATED BY
THE PAUL HAMLYN
FOUNDATION
TO THE
BRITISH MUSEUM

opened December 2000

TITLES IN THE SHIRE ARCHAEOLOGY SERIES
with their series numbers

Cover illustration
Bronze model of a ploughman and his team, from Piercebridge.
(Reproduced by courtesy of the Trustees of the British Museum.)

Published by
SHIRE PUBLICATIONS LTD
Cromwell House, Church Street, Princes Risborough,
Aylesbury, Bucks, HP17 9AJ, UK.

Series Editor: James Dyer

Copyright © Sian E. Rees, 1981
All rights reserved.
No part of this publication may be reproduced or transmitted in any form or by any means, electronic or mechanical, including photocopy, recording, or any information storage and retrieval system, without permission in writing from the publishers.

ISBN 0 85263 535 4

First published 1981

Printed in Great Britain by
C. I. Thomas & Sons (Haverfordwest) Ltd,
Press Buildings, Merlins Bridge, Haverfordwest.

THE BRITISH MUSEUM
THE PAUL HAMLYN LIBRARY
631.309361 REE

Contents

List of illustrations

Preface

This book is intended primarily as an introduction to the design and variety of prehistoric and Romano-British agricultural implements, in the most formative period for farming tools before the industrial revolution in Britain. It is also hoped that the book might act as a warning to the excavator, who, unless he is aware of the variety of designs and of possible functions of agricultural tools, may be tempted to use too easily the presence or absence of tools as evidence for specific activities on the site that he excavates. It would be rash, for example, to use the presence of a fragment of a curved iron blade as evidence of arable agriculture; the presence of querns on a site does not make it absolutely certain that the inhabitants grew grain themselves; the absence of iron ploughshares in an area during the iron age does not mean that the plough was not used. This book tries to show why.

The book is the result of study in museums in Great Britain. Agricultural tools from Ireland are mentioned if they are particularly good examples of rarer tool types. During the work, I received assistance from a great many people — friends, colleagues and museum curators throughout Britain. I would like to express my gratitude in particular to Mr P. S. Gelling, under whose guidance I originally became interested in the subject, and who kindly read through and commented on the text. I am also very grateful to Mr H. C. Bowen, Mr A. Fenton and Dr G. Hillman for their valuable assistance in the writing of the text and for their constructive criticisms of the more specialist aspects of the book. Thanks are due to the National Monuments Record of the Royal Commission on Historical Monuments (England), the Trustees of the British Museum, the National Museum of Wales and the Museum of English Rural Life, Reading, and to Mr P. S. Gelling for permission to reproduce photographs.

1
Introduction

By the end of the Roman period in Britain, all the agricultural implements that were used in Britain until the industrial revolution had been invented. Some of the forms of the tools were to be improved, and there was an increased emphasis on more durable materials as time went by, but the basic shape of each implement had by then been developed. The study of this formative period is very rewarding although it is hampered in parts by frustrating gaps in our understanding, usually due to the non-durable nature of the material from which tools were made. The study of the tools themselves and the changes in their shape and material is interesting in itself, but their study can be helpful in other ways. It can, for instance, increase our knowledge of the distribution within Britain of different methods of farming in different periods; also, taken with other evidence such as environmental evidence for the contemporary landscape, remains of plants and crops grown and animals raised and the remains of contemporary fields surrounding the site, surviving tools can help us interpret the agricultural activity of a particular excavated settlement site: whether, for example, it was arable or pastoral, specialised or mixed.

It is essential, however, not to misunderstand the nature of this evidence or misuse the tools for this purpose. Agricultural implements must be used alongside all these other sources of evidence, because the assessment of the functions of the tools is often far from clear-cut, as will be seen, and obviously a misinterpretation of the function of a tool will give an excavator the wrong idea about the site. Also, one of the great difficulties of studying a type of tool over a long period is that, inevitably, the same tool will be made of a variety of materials at different times and at different places. Some of these materials are more durable archaeologically than others, and, also, it often happens that a material was only used over a small area at a particular time. Hence, on the one hand, our knowledge of wooden tools is necessarily limited because wood is only preserved on archaeological sites in exceptional, usually waterlogged conditions; on the other hand, stone and iron for plough tips, for instance, were only used in specific areas at given times — stone in the second and early first millennium BC in the Northern Isles of Scotland, probably because of the lack there of hard wood used for toolmaking elsewhere, and iron in the pre-Roman iron age mainly in southern Britain, because this was where the earliest iron users in Britain happened to live. Therefore it is impossible to obtain a systematic overall view of the development and distribution of tools and the pattern is bound to be fuller at some periods and in some areas than others, because of this unevenness in the evidence.

2
Soil preparation: the plough and the field

The available evidence for the types of ploughs used in Britain in the prehistoric and Roman periods is largely but not entirely confined to the shares of ploughs. The share, the part of the plough which penetrates and undercuts the ground, is subjected to an enormous strain and hence tends to have been made of the hardest material available if it were suitable. Thus the stone, hard wood or iron of which the share was made has a better chance of survival than the presumably soft wood main body of the plough. Only a very few plough bodies survive from Britain but these, and two Roman bronze models of ploughs, necessarily schematic because of their small size, are all that we have to show us the shape of ploughs used in Britain in this period. Fortunately, a large number of plough bodies survive from north-west Europe because they were placed in peat bogs, in whose waterlogged conditions the wooden ploughs remain intact until, usually, modern peat cutting reveals them. As the wood of these ploughs is subjected to carbon-14 testing to date them we can gradually build up a picture of the types of ploughs used at different periods, particularly in Scandinavia, but also elsewhere in northern Europe, and compare it with the rather scantier evidence for Britain. In many cases we can assume that the share types for which we do have evidence would fit the plough bodies from northern Europe.

The types of ploughs

Four of these plough bodies, all from Denmark, are particularly interesting, as they are almost complete and between them represent the two main groups of early ploughs. The first three (figs. 1-2) — those from Døstrup, Hendriksmose and Donneruplund — are bow ards, or ploughs on which a share beam passes through a mortise, or hole, in the main beam. This allows an angled penetration of the ground. The plough from Vebbestrup is a crook ard, on which the main beam is mortised into the share beam or sole. This almost always means that the entry into the earth is horizontal (see fig. 3). All four are, strictly speaking, ards rather than ploughs: they are simple machines with no mouldboard or any device other than the share itself for turning the soil. Such simple ploughs have been given the name *ard,* a Scandinavian word, to differentiate them from the more sophisticated plough. Bow ards and crook ards are found in the Scandinavian and north European material, while in Mediterranean countries, though there is far less surviving material, the crook ard seems to have been dominant. There is no reason why both ard types should not have been used in Britain, though much of our evidence suggests that the bow ard was the dominant form. These two types of ard are illustrated with a later plough for comparison, and their component parts shown, on figs. 1, 2 and 3.

The Døstrup ard, found in north Jutland in 1884, consists of five parts: the beam, tie-hook, stilt, handle and foreshare. The main part of the ard, into which the other parts are mortised, is the 3 metres long (9 feet 10 inches) beam of alder with a hazel tie-hook mortised into the fore-end, to which the yoked oxen would be attached. The lime wood stilt has a handle attached at one end by which the ard would be guided by the ploughman, and on the other end it has a massive arrow-shaped ard-head. The stilt passed through a mortise in the beam and the ard-head helped to turn the soil. It has a long slot on its upper surface on which the foreshare, made from elder, rested. The foreshare, which acted as the cutting part of the ard, would have projected a little way in front of the ard-head and would have cut into the soil, which the arrow-shaped ard-head would then have been able to turn on to one side or the other if the ard was tilted. The foreshare is the type called a bar share. In this case the foreshare was pointed and worn on both ends, that is, it was reversible. The illustrations on fig. 1 show how the various parts of the ard fitted together. The Døstrup ard has been dated to 610 bc ± 100 by carbon-14 dating.

The more recent find of the Hendriksmose ard (fig. 2c and e), discovered during peat cutting in Viborg, Jutland, in 1957, is similar to the Døstrup ard and has a broadened ard-head in one piece with the stilt and a bar-shaped foreshare pointed and worn at one end. A notch has been cut in the upper side of the stilt near the back of the mortise, presumably for a wedge, now lost, which would prevent the foreshare from moving in the mortise. The beam has two notches at the traction end, which would have given two alternative positions for adjustments of the height of the beam, which would alter the depth of penetration of the earth. This ard, made entirely of oak, has been dated to 350 bc ± 100.

The Donneruplund ard (fig. 2a, b, d), found during peat cutting in central Jutland in 1944, is a more complex and sophisticated ard because of the addition of an arrow-shaped main share. It is basically similar to the ards described above but, instead of having a stilt which is widened to an ard-head to serve the function of a main share, it has two separate pieces. The stilt terminates as a small ard-head which widens just sufficiently to prevent the stilt from coming out of the mortise (fig. 2 b, 3), and the soil turning is performed by a separate arrow-shaped share (fig. 2 b, 2) which lies between the bar foreshare (fig. 2 b, 1) and the shorter ard-head. Thus the weight of the ard is reduced, but this system inevitably increased the instability of the ard, which was, in turn, reduced in three ways: firstly the notch cut on the rear end of the arrow-shaped share tang corresponded to a similar notch in the underside of the stilt, to which it was presumably secured by lashing; secondly, the two square holes on the top face of the arrow-shaped share originally held tenons to keep the foreshare in position (part of one of the tenons still survives); and thirdly, a notch on the upper side of the tang of the arrow-shaped share originally must have held a wedge to prevent the two shares moving in the beam mortise. Again, both ends of the wooden foreshare were worn and the arrow-shaped share has traces of wear on the point, particularly on the right-hand side, showing that the ard had been tilted to the right during ploughing. The beam of the ard is birch; the remaining parts are of oak.

The Vebbestrup ard (fig. 3a), found in north Jutland in 1929, is an example of a crook ard. A naturally forking piece of wood was used and the beam, in this case very short, only 136 centimetres (4 feet 5½ inches) long, was made from a birch branch, while a section of the trunk was used as the sole. A hazel wood stilt, only part of which was found, fitted into the sole to form the handle. With the ard was found an oaken swingletree, to which the short beam was obviously attached, and the end of the beam has three holes bored horizontally (shown by the finder's sketch, though only the rear one now survives). The swingletree would have been bound by a rope to one of the holes and the depth of the furrow could be regulated by moving the rope. The ard is dated to 910 bc ± 100.

The British plough: the wooden material

The only wooden part of an ard from Britain which is not part of a share or an ard-head is the Lochmaben beam (fig. 4b). It was found in 1870 in a peat bog near Lochmaben (Dumfries and Galloway). It is very similar to the beams from the Danish bow ards and is assumed to have been part of an ard of this type. The beam measures 248 centimetres (8 feet 1½ inches) in length, and the forepart of the beam is perforated by a rectangular opening for the attachment for the yoke. The rear part of the beam has a rectangular mortise shaped to hold shares at an acute angle to the soil. It is made of alder and is dated to 80 bc ± 100. In length, it lies between the Hendriksmose beam (*c* 2 metres, 6 feet 7 inches, long) and the Døstrup (*c* 3 metres, 9 feet 10 inches, long) and it is much longer than the Donneruplund beam (1.7 metres, 5 feet 7 inches, long). It is also less sophisticated than the Hendriksmose beam or the Vebbestrup beam and gave no alternative positions for the attachment to the yoke.

Three ard-heads and stilts have been found in Britain, one from Milton Loch (Dumfries and Galloway), dated to 400 bc ± 100, and two from a peat bog at Virdifield, Dunrossness (Shetland), undated, but similar to the Milton Loch implement (fig. 5). These ard-heads and stilts are very similar to those on the Døstrup and Hendriksmose ards and almost certainly fulfilled the same function on the same type of ard. They all have a roughly arrow-shaped head and a long straight shaft terminating in a rounded handle. The upper face of the head of the Milton Loch ard-head has two ridges and a deep groove between them which would have held a wooden bar foreshare, while the Virdifield heads have a rather shallower groove which would also have held a foreshare, which in this case might have been of stone. All three heads show signs of wear, particularly on the lower face at the tip, which must have borne the brunt of the force of the ard in the ground. They all also show asymmetric wear on the sides, showing that in each case the ards were tilted to one side or the other. The wings of the ard-heads must have been a great help in turning the soil.

The four remaining wooden plough parts from Britain are all ard shares. Two, dated to the first century BC and made of oak, were found in the waterlogged ditch of the rath at Walesland (Dyfed). The better preserved (fig. 6a and plate 1) shows a fire-hardened tip, which is evidence for an alternative method of making a wooden share more durable, other than by capping it with iron. This tool shows signs of asymmetric wear and these shares, now both broken, were probably the foreshares of bow ards. A similar though rather smaller section of a share, dating from the second century AD, was found in a well at the Roman legionary fortress at Usk (Gwent), and what is probably the remains of an arrow-shaped share, functioning as on the Donneruplund ard, has recently been found at Ashville, Abingdon (fig. 6b), dating to the third century AD. The implement is of oak, and the notches at its tip suggest that it had possibly been capped with iron to make the tip more hard-wearing.

The stone material

Thus the evidence of the wooden parts of ards remaining to us from British sites suggests that bow ards, similar to the Danish examples, were certainly used throughout the second half of the first millennium BC and the Roman period. Other evidence tends to reinforce this view and indeed extends the timescale over which we think this type of ard may have been used into the second millennium BC. There survive from many excavated sites in Orkney and Shetland examples of what are probably the stone shares of ards, or, more correctly, just the tips or points of the shares. A large number of such tools had been collected from the surface of the ground or during peat cutting near prehistoric sites in the nineteenth century and later excavations of such sites produced large numbers of rough stone tools of different types. So crude was their form and fashioning that the tools were often supposed to have been used for agriculture; they were made of stone because of the absence of hard wood in the islands. In the 1950s the ard points were recognised as distinct from the other tools and it was suggested that they were the tips of bow ards. They are nearly always made of sandstone local to the sites, pointed at one end, oval or round in cross-section, and are usually between 150 and 500 millimetres long (6 to 20 inches) (fig. 7 and plate 2). They are manufactured by firstly flaking a sandstone bar into the desired length and cross-section and then pecking the stone on the lower surface and sides, presumably to roughen the surface to prevent it from becoming smooth and from sliding through the mortise. Three shapes of butt or non-working end are found: double-pointed tools on which the butt end is pointed and the shape of the tools symmetrical and hence reversible; tapering-ended, on which the butt end is a rough squatter rounded end; and squared-ended, on which the butt end is a simple squared-off end usually at the widest part of the tool (fig. 8). Occasionally, the double-pointed tools are found worn at both ends like the bar foreshare of the Døstrup and Donneruplund ards.

One of the main reasons for thinking that the stone bars are ard points for bow ards is the type of wear marks that they display (fig. 7 and plates 3 and 4). When experiments with a replica of the Hendriksmose ard were carried out in Denmark, the wooden bar share was examined for the wear marks which the experimental ploughing had created; the marks, shown on fig. 4a, were very similar to those on the sandstone bars. The upper surface of the stone bars has longitudinal striations averaging 100 millimetres (4 inches) in length, but often more; the lower surface at the tip has a characteristic U-shaped wear, usually about 30 millimetres ($1\frac{1}{4}$ inches) long, and often stronger on one side than the other, showing how the tools were tilted during use; the sides also have wear, characteristic slanting striations, again often stronger on one side than the other and usually at about 28 to 30 degrees to the horizontal plane of the tool, showing the angle at which the tool had entered the ploughsoil. Occasionally the wear marks on the tools show that the shares had been turned upside down or, in the case of the double-pointed tools, back to front to stop the point of the tool, very worn from use, from becoming useless by being blunted. Turning the tools in this way made the stone cut the earth on a different side and hence at a sharper angle.

Although we can be fairly sure that we now understand the function of these sandstone bars, exactly how they were held in the ard remains unknown. From the position of the pecked surfaces we can guess where it was important that the stone should not slip and therefore presumably where the surface touched the wooden mortise, and we can also be fairly sure that the tools entered the soil at an angle. The points probably fitted into a wooden mortise of some kind, either in a composite separate wooden share, or more probably in the share beam itself. The tools are usually associated with the oval and heel-shaped and less commonly the round houses of the second and early first millennium in the Northern Isles, but they have also been found in burnt mounds, cairns, earth houses and even brochs; this suggests that they were still in use later in the first millennium.

The iron material

From the iron age onwards in Britain, iron shares were made for attachment to ards. The largest proportion of all iron shares remaining to us are again the tips of foreshares rather than the foreshares themselves. They are small iron shafts pointed at one end and with a socket or encircling flanges at the other to hold the end of the wooden part of the foreshare. They vary considerably in length, from the shortest sheath, which would fit over the very end of the wooden share, only 70 millimetres long ($2\frac{3}{4}$ inches), as with the iron age example from Gussage All Saints, to a longer flanged bar covering a substantial amount of the wooden share, such as that from Bigbury dating from the mid first century AD (fig. 9). Some have holes for nails to give additional attachment to the wood, such as the example from Brading (Isle of Wight). The final development of this process of lengthening the amount of iron used in the foreshares was to make the whole share of iron. Some of these all-iron bar shares survive from fourth-century AD contexts from Roman Britain, from the hoards from Dorchester (Oxfordshire) (one), Great Chesterford (Essex) (five), Silchester (Hampshire) (six), Worlington (Essex) (five) and Abington Pigotts (Cambridgeshire) (one) (fig. 10). In every case save that of the Worlington hoard the shares were found with an equivalent number of coulters. This fact and the wear on their points led to their identification as shares. They were presumably held in a similar way to the wooden foreshares of the Danish bow ards. These fourth-century bar shares are usually between 60 and 90 centimetres in length (2-3 feet), octagonal or square in cross-section and have a flat chisel-like point hammered out from the bottom and sides only of the iron. The smaller iron sheaths very rarely show any signs of wear because their slighter bodies probably could not have stood up to as much use as the later, stouter tools, and they tend to be broken or very corroded. The longer bar shares have survived better and a number show wear on their pointed ends of the same type as the stone and wooden shares described before, often with asymmetric wearing on one side. One of the Silchester shares was double-pointed and hence could have been reversed. Most, however, terminate in a blunted knob.

As well as the bar shares of iron, there are other shares surviving from the Roman period in Britain. Firstly, there are the broad spade-shaped tools with wide curving ends, often with the flange continuing all around the sides and end of the pointed share. These broad tools are from 10 to 17 centimetres (4-$6\frac{3}{4}$ inches) long, 7.5 to 12 centimetres (3-$4\frac{3}{4}$ inches) in width, and are vastly heavier than the sheath-type shares. They are not common and are not found before the Roman period. Fig. 11c shows the example from Blackburn Mill, which is first or early second century AD and has a nail hole in the back. Secondly, there are the symmetrically flanged type of shares with stout triangular blades and a large flat open socket, oval in section. Again, these are only found dating from the Roman period onwards, but from fairly early in the Roman period. Fig. 11b shows a good example from Bucklersbury House, London, which dates to the first or second century AD. Another, from Frindsbury (Kent), may have been associated with a coulter — the records of the excavation do not make it clear but in Rochester Museum, where they are now, the two have always been regarded as being associated. Related to these symmetrical flanged shares are the winged flanged shares, which are roughly the shape of a right-angled triangle with the blade edge as the hypotenuse (fig. 11a). These, like the symmetrical flanged shares, are massive with large heavy open flanged sockets. One comes from the Roman villa at Folkestone, one from the villa at Brading (third or fourth century), another from the North Welsh hillfort at Dinorben (probably early fourth century) and the fourth comes from Chester. Two have the wing on the right, two on the left. They are good evidence that by this time the plough with the fixed mouldboard had been invented, since they would cut the earth on one side only, and the flexibility of the other shares, which could be tilted one way or the other, had been abandoned. This suggests that the plough had been equipped with a turning agent, the mouldboard. Finally, there survives one massive and most unusual share from the villa at Box, presumably, though not absolutely certainly, Roman in date. How this tool fitted on to a plough is unclear, but it seems likely that it and the other heavier shares — the spade-like, the symmetrically flanged and the winged shares of the Roman period — were fitted on to a horizontal sole. This in turn suggests that a different type of plough may have existed alongside the Danish-type bow ard in the Roman period — possibly a bow ard with a horizontal share beam on to which the share was slotted, perhaps a type of crook ard with a sole beam on which the share was slotted (fig. 3).

Coulters of iron are not uncommon finds from Romano-British contexts. Fourteen have been found in hoards of Roman ironwork, seven come from excavated sites and six are stray finds. Many of these are associated with bar shares, showing that the two were used together presumably on a bow ard. One is possibly associated with a flanged symmetrical share and we would certainly expect them to have been used both with these and with the winged shares on a heavy plough. The coulter is a knife-like implement which, when attached to the plough, cuts the earth vertically (fig. 3). The share follows and cuts the earth horizontally and the mouldboard inverts the soil thus cut. Coulters from Romano-British contexts are fairly standard in shape and consist of a triangular blade, with the blade edge on the longer slanting side, and a long shaft, usually octagonal in cross-section, ending in a hammered knob (fig. 10). Most of the coulters which survive are fourth-century in date, though this may be because hoards of large iron tools are more common from the fourth century than from earlier, rather than because of the date of the first or commonest usage of the tools.

Evidence for the techniques of ploughing

The Piercebridge plough model of bronze (fig. 12), found at the Roman fort at Piercebridge (County Durham) and probably second or third century AD in date, shows a plough in use. We have to be careful in using this model as evidence, firstly because it is so small and the plough necessarily rather schematic, and secondly because it may well have been a model of a ritual rather than a normal utilitarian ploughing scene. It seems to portray two oxen yoked to an ard giving a horizontal passage through the ground. In the beam in front of the share is a small round hole, which was probably for a detachable coulter, now missing. On either side of the junction of beam and stilt is a pair of forward curving arms and Manning suggests these were supports for detachable earthboards. These boards were described by classical authors (Varro 1.29.2; Pliny *NH* 18.180) as being used to cover broadcast seed and to cut ditches for drainage or for clod breaking after sowing, in effect a kind of proto-mouldboard. Without the help of these comments from the classical authors, we would not have been able to interpret these arms. The Piercebridge plough model shows the right hand of the ploughman as being thrust forward, and Manning interprets the piece of metal which projects back from the palm as the remains of a long ox goad with which he was originally urging the beasts forward. We know that the ox gold was used, as many continental mosaics show ploughing scenes with the goad being used, though Columella advises against it as he says it makes the oxen 'irritable and inclined to kick'. Many iron points with a socket at one end and a spike at the other survive from iron age and Roman contexts and these are often interpreted as ox goads, though the interpretation of such basic iron sheaths is inevitably open to some doubt.

Roman mosaics suggest that oxen were the normal draught animals used and engravings on rock outcrops in Scandinavia, probably datable to the bronze age, suggest that even at this date teams of two or four oxen yoked together were the normal method of traction (fig. 12b). Two types of yoke — the head or horn yoke, attached to the horns of the oxen, and the withers yoke, attached to their shoulders — seem to have been used in Britain in the pre-medieval periods, and examples of both kinds have been found in peat bogs in Scotland and Ireland. The head yokes have lightly curved neck pieces and horizontal central openings for the thongs which fastened them to the horns of the animals (fig. 13), while the withers yokes have more deeply curved side pieces and central horizontal openings to hold lashings for a beam or pole but usually also have vertically bored holes at the side of the neck piece for thongs which encircled the animals' necks. Both Columella and Pliny advocate the use of the withers yoke and condemn the head yoke as inefficient and cruel as the oxen are so uncomfortable with their heads pulled back that only light ploughing can be done.

The fields and ploughing

Now that we have reviewed the evidence for the sort of ploughs that were used in prehistoric and Roman Britain, we should look at how and where they were used. Evidence for this comes from the excavation of the marks that the ploughs left in the ground, the shape and size of the contemporary fields and modern experiments with plough replicas. Since the 1950s an increasing body of evidence from the excavation of ard marks has taught us a lot about ploughing in the prehistoric and Roman periods in Britain. Two patterns of ard traces are frequently discovered: those which form a criss-cross grid, and those which run in parallel single-direction grooves. It has been assumed that criss-cross ard traces were created by ards without a mouldboard, cross-ploughing in order to loosen the soil thoroughly. Single-direction traces were probably made by a heavier plough which could turn the soil and therefore did not need to replough in a different direction. The traces were created when the point of the share penetrated the subsoil, making a groove into which the ploughsoil fell. When the subsoil is of a markedly different colour from the ploughsoil, for example dark brown ploughsoil on a yellow sand, the traces show up very clearly (plate 7). Criss-cross ard marks excavated on a variety of sites have shown us that the light ard could cope with a range of soils including heavier soils such as clay. From the cross-section of the ard traces, often an asymmetrical V-shape, we may gain some impression of the degree of tilt at which the ard was commonly held (plate 8 and fig. 3d). A large variety of angles is found, as is to be expected with something as crude as holding a plough at a tilt despite irregularity in slope, the stone content of the ground and the smoothness of the draught. Any angle up to 44 degrees from the vertical has been found, while up to 20 degrees is very common. Ard traces are often found bunched at the sides of fields and this suggests that the edges of the fields were ploughed over to tidy the furrow ends and reduce the weed-infested uncultivated parts. It is frequently noticeable that an ard trace does not turn aside at an obstacle such as a large earthfast stone; this suggests perhaps that the plough teams were led. Criss-cross ard traces have been found under long barrows of the neolithic period and in a variety of contexts throughout the prehistoric periods. By the Roman period, however, it is more common to find single-direction parallel ard traces probably created by the heavier Roman plough.

It would be foolish, however, to suggest that these ard-trace patterns are completely understood. The traces that we find seem frequently more widely spaced than would be compatible with achieving a proper tilth, and also any ploughing technique which always produced ard traces would, after a few years, create a confusion of traces rather than the grid patterns that we often encounter on excavated sites. It is possible that the ploughing that created the ard traces was a special deep ploughing, for breaking a fallow perhaps, or even for levelling an area for a new field.

The relationship between ploughing practice and field shape is obvious, and we shall now look briefly at the field types in use in the prehistoric and Roman periods in Britain. The so called 'Celtic' fields (plate 5) are fairly clearly associated with the practice of cross-ploughing (criss-cross ard traces were found in the Celtic field system at Overton Down, Wiltshire). Traces of these square or rectangular fields are found along the south of Britain with lesser survivals in the non-acid soils of the Midlands and northern England, and recent surveys are leading to further discoveries. Celtic fields have been found on ground that is virtually flat, defined by banks, baulks or slight lynchets. Their clearest remains, however, are as lynchets, which, on steep slopes, may be 2 metres (6 feet 6 inches) or more high. Such big lynchets are difficult to eradicate from aerial view but are all too easily spread by modern levelling and in a relatively short time they can disappear totally, since they cannot (unless by following a ditched line) be recovered by excavation. Concentrations of flint or other stone gathered off the fields are frequently found in lynchets, while some lynchets mask stone walls. It is also often unclear whether the curvilinear enclosures of quite a different type from the Celtic fields (plate 6) were for arable or pastoral farming, so we have only a very incomplete picture of arable field systems and their distribution in prehistoric Britain. However, within this limitation, the study of field systems can tell us a good deal about prehistoric farming. The earliest of the Celtic fields may date from the early bronze age and, as these are permanent field systems deliberately laid out, this implies that farmers had come to some understanding of soil properties and had worked out some policy for soil regeneration, presumably by leaving fallow, manuring and crop rotation. Bowen has recognised two basic types: small rectangular fields generally about $\frac{1}{3}$ to 1 acre (0.1 to 0.4 hectares) in size and rarely more than twice as long as broad; and longer narrow fields (with 'closed' square ends unlike medieval strips) up to six or so times as long as broad. The longer fields in places seem to have replaced the small square fields and Bowen suggests that the long rectangular fields may have been due to the introduction of the heavy plough, which did not need to cross-plough and which would have been more difficult to turn. Lynchets are the result of plough action as may be the low baulks sometimes dividing fields on the flat or up-and-down slopes. Fields were originally marked out in some other way, which included dwarf walls and slight banks and ditches.

Experiments with ploughing

Experiments with replicas of prehistoric ards can also tell us something of ancient ploughing techniques in Britain. The experiments of Aberg and Bowen with a replica of the Donneruplund ard (plates 9-12) and those of Hansen with a replica of the Hendriksmose ard suggested that these ards were not at all efficient at breaking up fallow or even hard dry soil under stubble. The ard slid or jumped repeatedly on the hard ground. Also Hansen found it difficult to tilt the ard in loose soil as the ard tended to slip sideways. His experiments used two oxen for draught, yoked with a replica of the Dejbjerg withers yoke, and provided some useful information about the best position for the foreshare (which was for it to jut out at least 100 millimetres, 4 inches, in front of the ard-head in order to achieve a better cut of the ground) and about the speed and efficiency of ploughing with this ard (he found that the walking speed of the oxen during ploughing was between 3.6 and 4.6 kilometres, $2\frac{1}{4}$ and $2\frac{7}{8}$ miles, an hour). Both the experiments of Aberg and Bowen and those of Hansen showed the high rate of wastage of the foreshares as they wore very fast or broke. Some allowance always has to be made with such experiments for the relative inexperience of the experimenters. Hansen's experiments also created in the subsoil criss-cross ard traces, which are useful comparisons with excavated examples from prehistoric contexts.

3

Care of the soil and crop : hoes, spades and weeding tools

The prehistoric period

Other implements were needed as well as the plough both for the initial tillage of the soil and then for the tending of the crops. Unfortunately, our knowledge of prehistoric manual tillage tools is very slight. Occasional claims are made that stone or bronze axes were used in ground clearance and the wear marks and roughness of some of the stone tools from prehistoric sites in Orkney and Shetland do suggest that they may well have been used for ground clearance. It has also been suggested that neolithic stone 'mace-heads' might have been used as weights for digging sticks and that the large stone 'axe-hammers' of the bronze age may have had an agricultural function. However, the case for such implements is not proven and it seems likely that early manual cultivating tools would usually have been made of wood, which has not survived, or that the seemingly unspecialised antler hoe was used. This implement is found in a variety of contexts, in flint mines, for example, or in the ditches of barrows, both situations implying a non-agricultural function. A more specialised tool, the perforated antler hoe, is an uncommon find from bronze age contexts. It has a round perforation for a handle and a chisel-shaped end, which often bears wear marks compatible with use as a light hoe (fig. 14d). A few wooden spades have been found on prehistoric sites in Britain but their specifically agricultural function can rarely be proved. For example, a probably iron age spade with a rectangular blade and handle from a site in Skye may have been an industrial tool, to judge from its context, and it is uncertain whether the neolithic long-handled wooden object from Ehenside Tarn (Cumbria) is a spade or a paddle. Surviving wooden spades from prehistoric contexts are in any case very rare, though the excavation of D-shaped spade marks around a prehistoric field at Gwithian (Cornwall) suggests that here a wooden spade with a heart-shaped blade was used to tidy up the unploughed periphery of the field after the rest of the field had been ploughed.

It is not until the Roman period in Britain that the commoner use of iron makes agricultural tool types more apparent to the archaeologist; in this period, however, there are quite a variety of tools for the manual tillage of the ground. Four types of iron hoe, iron mattocks and a great variety of shapes of iron spade shoes, as well as weeding tools and turf cutters, have survived for our study.

The Roman hoe

The first type of hoe for which we have evidence is the entrenching tool, though, to judge from its predominantly military distribution in Britain, this was a primarily military tool (fig. 15a). Entrenching tools have a wide spade-shaped blade on one side of the head and a pick blade on the other. Both blades are held at an angle of slightly under 90 degrees to the wooden handle, a feature that is normal on all hoes and mattocks, Roman or modern, to make it easier for the blade to bite into the ground. The spade-shaped blades are usually about 17 centimetres ($6\frac{3}{4}$ inches) long, while the pick blades are usually slightly shorter. These tools are also very common on military Roman sites in continental Europe. They were quite an early introduction into Britain and appear in contexts dating throughout the Roman occupation on military sites all over Britain.

The second type of hoe is the single-bladed type, which again is common on Roman sites in Britain and Europe, though it is difficult to be sure whether some of the smaller tools are hoes or adzes. The larger wider-bladed tools are almost certainly agricultural and a variety of blade shapes is found. Two examples can be seen illustrated on fig. 16 b, c. The third type of hoe is a smaller tool with a short stout blade on one side of the head and two tines on the other (fig. 16 a, d, e). K. D. White, in his book *Agricultural Implements of the Roman World,* suggests that this may be the *ascia-rastrum,* a tool described by the classical agronomists as being used to aerate the soil, remove weeds and tend young plants on a more delicate scale than would be possible with any of the tools hitherto described. The tines would be effective in loosening the soil and the blade would draw it. Few are well dated but they seem to have been a fairly early introduction into Britain in the Roman period. They are usually about 20 centimetres (8 inches) long, and a variety of blade types is found. Again, they are quite common finds on Roman sites in Britain and Europe. The final iron hoe type is the *bidens,* the name used by the classical agronomists to describe a two-tined hoe and attributed by White to this tool type (fig. 15 b-d). These are rarer tools and only seven are known from Britain. They are heavy two-tined tools which must have been used for soil breaking. Some have sockets while others have perforations for a wooden handle, and, like the other hoe types, they seem to have been a fairly early introduction into Britain. Possibly related to these tools are two-tined antler objects, which are not infrequently found on Roman sites, though they are generally rather smaller and presumably would not have stood up to such heavy work (fig. 14). The function of these antler tools is, however, by no means certainly understood. Some of the stouter ones were probably hoes, while others have more delicate tines and would have been unsuitable for such work.

In addition to these hoes, there survive straightforward mattocks, which are not solely agricultural tools, though their broad blades (fig. 17) are quite suited for clearing rough ground and breaking clods, while the axe blades could be used for cutting off old stumps and shoots, and it is clear from the classical writers that they were indeed used for such functions. Most mattocks have two blades: a broad blade which would cut square on to the earth; and an axe blade, the cutting edge of which lies at right angles to the mattock blade. The pickaxe, on the other hand, has an axe blade on one side of the head and a slender pick blade on the other. Pickaxes would have been used more in military or forestry operations. Mattocks are quite common finds on Roman sites in Britain and Europe and seem to have been used throughout the Roman period. They are often about 27 centimetres (10½ inches) long, though some are considerably smaller.

The Roman spade

A very common find on Romano-British sites is the spade shoe, which protected the wooden blade from excessive wear by encasing the cutting edge and sides in an iron sheath. 'Spade shoe' has become the usual archaeological term for these objects though, strictly, they are likely to have been sheaths for shovels rather than spades. Over forty sites in Britain have produced more than seventy examples of spade shoes and their variety of shape and size and of the methods used in attaching the iron to the wood is remarkable. Unfortunately, as far as we know, no particular size, shape or method of attachment is specifically Roman and, as the tool continued in use in Britain until comparatively modern times, this makes their typological dating impossible. Also, few of the different types seem to have any specific area of distribution or chronological importance, so the different types will just be noted briefly (figs. 18-19). There are round-bottomed and square-bottomed shoes, which would have fitted the equivalently shaped wooden spade. Round-bottomed spade shoes can either be a simple curved sheath (fig. 18c) or have flaring arms and a blade which flares to give a wider cutting edge of iron than there is on the wooden spade (fig. 18d). Another type has a flaring blade, upright arms and lugs to grip the side of the wooden spade, while another has a simple curved blade with straight arms which then run along the top of the wooden blade. The square-bottomed shoes (figs. 18e, 19) have a similar variety of types. The range of types, findspots and dates is large. Spade shoes are found on villas, town sites, forts and small settlements and their date range stretches from early in the Roman period onward. When small parts of shoes are found, it is often difficult to show to what type they belong.

The variety of site types suggests, as one would expect, that they were unspecialised digging or soil-lifting tools and not purely agricultural in function. A few wooden spades dating from the Roman period have been found in Britain and one round-bottomed spade from Silchester (fig. 18a) still bears the marks of the iron shoe which it once wore. A fine rectangular wooden spade, about 40 centimetres (16 inches) long, with a diamond-shaped tang, survives from Chester (fig. 18b) and as it was worn asymmetrically along the cutting edge it was probably never shod with an iron shoe. Iron shoes come from predominantly Romanised areas and it is probable that in other less highly Romanised areas the wooden spade was generally unshod.

Some all-iron spades have been found on Romano-British sites. One large unwieldy shovel was found at the hillfort at Croft Ambrey (Hereford and Worcester) but more common are small flanged rectangular blades which have been described as ploughshares in the past but which may well be peat spades (fig. 20). It was first suggested that they might be peat spades as two very similar tools were found in a first-century or early second-century hoard of Roman ironwork from Blackburn Mill (Borders). One of the tools had a footpiece and was clearly a peat spade (fig. 20b) and it was then suggested that a group of similar blades from a variety of Roman sites in England and Scotland were peat spades also. They would have shod a wooden spade which would have been fairly narrow, but could have been quite long, like the modern peat spade. One of these tools, that from Brampton, has a nail hole at the back of the socket to assist its fastening but the blades, like the iron shares, would presumably have been fitted by being heated and then shrunk on to the wooden blade.

Weeding tools and turf cutters

A reasonably common find on Romano-British sites is the wide-bladed socketed tool often described as a spud or weeding tool. This interpretation is quite plausible. The tools are usually socketed (fig. 21e) and have short blades with which delicate work around plants could have been carried out. Rather less common is the iron double-winged implement so similar to the present-day turf cutter that its identification as such seems certain. Both these tools seem to have been introduced into Britain quite early in the Roman period and have a fairly wide distribution on a variety of sites. The turf cutter was probably a military introduction used to cut turf for rampart construction. Romano-British turf cutters may be socketed or tanged (fig. 21) but one tool from Great Chesterford (fig. 21c) is unique. It has a triangular blade and a foot rest and is considerably longer than the other more orthodox examples with their curving blades.

4

The harvest: sickles, hooks and scythes, pitchforks and rakes

The harvesting tool is one of the earliest of man's agricultural tools for which we have evidence, and sickles made from a variety of materials have been found dating from the early neolithic, particularly in the Near East: sickles of burnt clay were found in early levels at Eridu in Mesopotamia, while Twelfth Dynasty levels in Egypt produced composite angled flint sickles. The main reason for this comparative wealth of evidence is that for a tool to be sharp enough to cut grain stalks it cannot be made of wood or antler. Flint was the first material to be used in Britain as far as we know and it is very durable archaeologically. During the bronze age bronze was used for sickles and later sickles were made from iron; perhaps partly because of the necessity of having a fair number of such tools and partly because such small tools did not require a large quantity of iron, therefore reducing their scrap value, we do have a comparatively large number of iron harvesting tools remaining to us for our study.

The prehistoric reaping hook of flint

Steensberg, in his important work *Ancient Harvesting Implements*, refers to early harvesting tools as being of three different sorts: the straight reaping knife, the angular reaping hook on which the angle between the blade and handle is acute, and the balanced sickle on which the blade is bent rearward from the handle to provide a more balanced and convenient implement. Examples of flint, probably of the first and second categories, are found in the neolithic period in Britain. The so called 'crescentic flint sickles', which probably date from the later neolithic period, were probably hand-held and hence should be described as reaping knives (fig. 22a). The main reason for their identification as reaping tools is their shape and similarity to more finely made crescentic sickles from Scandinavia. The small non-crescentic sickle flints are a different sort of tool. They are often called sickle flints because on their surface can be seen a gloss, which can be achieved by cutting straw containing silica (fig. 22c). The flints are very variously shaped and it is only by this gloss that their interpretation is postulated. However, it is becoming clear that such a gloss can be obtained by the cutting of other materials and it is only by a very time-consuming and as yet young technique of micro-wear analysis, looking at the gloss through magnification, that proper identification of the function of an individual flint can be made. Hence, while it seems likely that small sickle flints would have been set in a wooden groove in an angular composite hook, we cannot be certain that this was so, and we are still less certain what shape the entire tool would have been. In Denmark a tool with a single large flint set at right angles to the handle was found with the original wooden handle at Stenild, and it has been suggested that this was a weeding tool rather than a sickle (fig. 22d, 2). Though similar shaped flints have not to my knowledge been found in Britain, a variety of flint-bladed cutting tools was used and we shall have to do a great deal more work on wear analysis before we have a better idea of the variety and function of these tools. It is possible to harvest corn without a harvesting tool, by plucking it by the roots. This practice still occurs in parts of the world, the Greek island of Thera for example, and was known within living memory in the Hebrides. This type of harvesting would leave no archaeological record. There also survive from Shetland a few serrated slate hooks, which may have been used as sickles (fig. 22b), as slate could have given a sufficiently sharp cutting edge for this purpose. These date from the bronze age but have so far only been found in Shetland and were not in general use in prehistoric Britain.

Bronze hooks

Bronze reaping hooks survive in considerable numbers, mostly from bronze hoards. There are two different types found in Britain: the socketed tools, far commoner in Britain than in mainland Europe, and the non-socketed tools, either those with a knob or those with a tang and rivet, which are the commonest continental types. The earliest of the socketed sickles date from the end of the middle bronze age and consist of a long straight double-edged blade set laterally to a cylindrical socket open at both ends (fig. 23a). Later, the sockets of the sickles become conical with the upper hole becoming smaller and the blade rising towards the top of the socket, which becomes more oval in cross-section. The next development is for the upper hole of the socket to become covered over (fig. 23b) and finally the blade edge continues over the top of the socket (fig. 23d). Alongside these sickles are vertically socketed sickles which seem to be derived from the late bronze age socketed knife. Their only development can be seen in the degree of curvature of the blades: early tools such as that from Rosebury Topping (fig. 23e) have straight blades while later tools like that from the Thames at London have more strongly curving blades (fig. 23f). All these vertically socketed tools can be dated to within the later bronze age.

The two types of non-socketed sickle are rather different, not only in their method of hafting but also in that their blades are single-edged only and usually are cast plate, i.e. they are not symmetrical in cross-section like the socketed sickles (fig. 23 g-i). The vertically tanged sickles (fig. 23g) provide a rectangular tang with side ribs and a rivet to attach the wooden handle. The knobbed sickles may have one or two round knobs (fig. 23h) or one elongated knob (fig. 23i) to attach the handle, and all three types are found on sickles from Britain. All the tanged sickles are datable to the late bronze age, but some of the double-knobbed sickles date from the later part of the middle bronze age. One of the most interesting of the hoards in which bronze sickles have been found is that from Llyn Fawr (Glamorgan) (fig. 23d and plate 13), which must date from the very end of the late bronze age. Three very similarly shaped socketed sickles were found in this hoard, but one was made of iron.

The socketed and the tanged sickles were hafted so that the main part of the blade lay at roughly 90 degrees to the handle. It is less clear how the straighter, knobbed sickles were hafted. Their handles could either have been in the same plane as the blade so that they would have been reaping knives, or they could have been at right angles to the blade. Steensberg carried out some experiments with bronze sickles and used a knobbed sickle hafted at right angles. He found that the tool worked reasonably well as a reaping hook and took a similar time to cut corn as the crescentic flint sickles, but it was slightly more accurate and uprooted fewer plants. The sickle blades were sharpened on an anvil with a hammer, traces of which process are often still visible on bronze blade edges. The variety of shape of the tools suggests that they may well have had a variety of functions, cutting different materials for different purposes. Because of the limited distribution of bronze sickles in Britain, the costliness of manufacture for so utilitarian a tool and the fact that they are sometimes found with decoration, it seems probable that flint sickles continued in use throughout the bronze age alongside the bronze tools.

Iron hooks and sickles

A large number of small iron tools with a concave cutting edge and socketed or tanged handle attachment have been found on iron age and Roman sites in Britain. Any attempt to work out their function is bound to be somewhat suspect because of the enormous range of sizes and shapes encountered, especially if we consider the great variety of tasks to which curved iron tools could be put on a prehistoric farm: the gathering of roofing, flooring and other building materials and materials for basketry are examples of non-agricultural tasks, not to mention more specifically agricultural functions, such as weed clearance, fodder gathering, pruning, harvesting and so on. However, there are two obvious categories into which most tools might at least loosely fall: the larger tools could have been used for harvesting or weed clearance, while the smaller hooks would be more appropriate for the tending of small trees and bushes, and the exclusive range of blade shapes found on the tools of these two groups does support the independent existence of the two groups of hooks. Both the larger tools, described here as ‘reaping hooks’, and the smaller ‘pruning hooks’ can be socketed or tanged.

A group of large socketed reaping hooks are similar in shape to the late bronze age laterally socketed bronze sickles (see the two examples from Llyn Fawr in plate 13 and fig. 23d and compare with the Roman tool from Codford St Mary in fig. 24a), and this is a long-lived tool type, which survived well into the Roman period. This group, because the shape of the tools co-existed for a long time with the efficient balanced sickle, had most probably a specialised function not necessarily anything to do with the harvest. A second group, the most commonly encountered tools, with simple curving blades and socketed or tanged methods of hafting, are perhaps more likely to have served as reaping hooks (fig. 24 b-e), though these too continued well into the Roman period. Finally, the balanced sickle, which seems to have been introduced from Europe in the later part of the iron age, appears first in hoards such as the Llyn Cerrig Bach hoard (plate 14) and at Glastonbury (fig. 24 f-h) and continued in use throughout the Roman period, becoming the normal harvesting tool in medieval Britain. Other hook types with longer blades set at right angles to the handle do exist, tanged or more rarely socketed, and can be balanced or non-balanced, but most of these are stray finds and can rarely be given a definite Roman date. The commonest reaping hooks are those three types described above.

The classical agronomists provide us with information about implements available for grain harvesting, though they are usually describing the Mediterranean and not Britain. Particularly interesting is Varro's description of three methods of harvesting known to him. The first, used at Umbria, was to cut the grain at ground level with the sickle and then cut the ears off sheaf by sheaf. The second, used in Picenum, was to cut bundles of plants close to the ear with a curved piece of wood with an iron saw at the end and subsequently to cut the stalks close to the ground. The third method, used near Rome, was to cut the plant halfway down the stalk, leaving half of the length of the stalk to be cut later. The method used depended on whether the straw was wanted for bedding or thatching or if it was wanted at all. Small curving iron saws have been found on Romano-British sites and may have been used for harvesting, as described by Varro, or for pruning.

Pliny and Palladius both make interesting references to a harvesting machine or *vallus* which was apparently used on the large estates in Gaul. This machine is illustrated on a number of Romano-Gallic stone monuments (fig. 25), so that we know there were at least two different sorts of machine. Basically, both types consisted of a row of teeth which would strip the ears so that they would then fall into a container. The wheeled machine was pushed by a donkey or an ox, and an operator supervised the filling of the hopper and kept the teeth free from chaff. An experiment with a replica of a *vallus* took place at Virton, Belgium, in 1969, when it was found to work satisfactorily. No remains of the *vallus* survive, and we cannot know if it was ever used in Britain.

The smaller group of tools, the pruning hooks of iron, are also quite commonly found on iron age and Roman sites in Britain. By far the greater number are socketed rather than tanged and four blade shapes seem to have been used — curving, hooked, angular and upright (fig. 26) — and, with their consequently different-shaped cutting edges, it is probable that a great variety of functions were carried out by the tools. The socketed tools, like the reaping hooks, can have open or closed sockets and often have a rivet hole in the back or two holes in the sides for a cross rivet to fasten the wooden handle. Tanged tools have either spikes or flat tangs for handle attachment.

Billhooks

As pruning hooks were designed for small-scale delicate work on plants and bushes, so billhooks were for larger-scale chopping work on trees and hedges, and their main function may well have been the gathering of winter fodder for animals. They had acquired their distinctive modern form by the end of the iron age. They are larger and more solid than sickles and reaping hooks and have broad blades with the main section of the cutting edge in the same plane as the handle. The most commonly found tools (fig. 27a) have a beak on the end of the blade, which would be used for drawing the branch towards the operator so that he could hold it with one hand and chop it with the tool in the other. They are usually socketed but are occasionally tanged and are found in iron age and Roman contexts. The second type of tool has the blade lying in one plane only but often has a talon on the top of the back of the blade. These are rather less common and are always socketed but again can be iron age or Roman (fig. 27b). The third group of bills consists of long socketed choppers with large curving blades set at about 140-160 degrees to the socket (fig. 27c). They often have a projection at the tip of the blade, probably fulfilling the same function as the beak on the blades of the first group. They all seem to be Roman in date, as is the final group of bills, which has axe-shaped blades and long socketed handles (fig. 27d). Neither of the last two groups is as commonly found as the first two. Another possible type of bill is the heavy tanged tool which used to be known as a coulter but which was reinterpreted as a billhook because of the dissimilarity to all other coulters and the seeming impracticability for such a function. However, the two known tools of this type are so much heavier than the ordinary billhooks and so clumsy and unwieldy, as well as having mysterious attachments at the bottom of the tangs, that this reinterpretation also seems unlikely. One of the tools comes from Bigbury (fig. 28c) and would have a mid first-century AD date; the other comes from Wroxeter and is Roman in date.

Bigbury hillfort has produced another group of very interesting tools quite unlike any other cutting implements from Britain. Again they presumably date from the mid first century AD and the five tools, all with open sockets, save one which is tanged, are probably best regarded as slashing tools for ground clearance (fig. 28a, b).

They have long wide blades, which, on two of the examples, have a rolled round strengthening rib on the blade back, a feature otherwise characteristic of scythes. The tools do have features akin to scythes and this led to a theory that they were a sort of early scythe with long handles turning through 90 degrees to allow the operator to mow with them while he stood upright. However, the tools would have been very unwieldy if hafted in such a way, and their shape seems best designed for use with a short handle for slashing. It is strange, however, that they are so uncommon a tool type.

Scythes

The discovery of scythes or fragments of scythes on Romano-British sites is not uncommon and a surprising variety of scythes survives from this period. The commonest scythes are those with the tang set at an acute angle to the blade: they are balanced scythes with the elbow of the blade on the one side of the handle offsetting to some extent the weight of the blade on the other side. Within this type there are two distinct subgroups: the shorter scythes, which vary from 84 to 120 centimetres (33-47 inches) overall length; and the longer scythes, all late in date, which are similar in shape but are far longer, ranging from 130 to 160 centimetres (51-63 inches) overall length. The smaller scythes were introduced into Britain quite early in the Roman period and a group of four found in a pit in the Roman fort at Newstead date to the first century AD. They were quite probably a military introduction. The Newstead tools (fig. 29e) have strongly curving wide blades with a strengthening rib on the back edge and have rivet holes at the junction between blade and tang for attachment to a long wooden handle. This smaller tool continued in use throughout the Roman period; others of the same basic shape had slimmer blades than the Newstead tools.

The longer scythes have a basically similar shape, but they are far longer and have slimmer, less curving blades (fig. 29 a-d). These scythes have always been a focus of particular interest since their original discovery in a hoard at Great Chesterford (Essex). Since then, others have been found at Abington Pigotts (Cambridgeshire), Farmoor and Hardwick (Oxfordshire) and Barnsley Park (Gloucestershire). Those which are datable belong to the fourth century. They are so extraordinarily long that in 1967 the Museum of English Rural Life in Reading made a reconstruction of one of the Great Chesterford scythes and experiments were carried out to test the practicability of the scythe in use (plate 15). Various types of snead or wooden handle were used, all 'shoe-shaped' at the lower end to facilitate clamping on to the tang by means of two iron rings, each tightened by a small wedge under the tang. The scythe was used to cut a thick crop of tall grass and crops of modern winter wheat and barley. An experienced mower used the tool with little bother, using either a full swing with a wide cut with the mower moving forward, or short chopping strokes with the mower moving sideways. It is not certain whether the scythes were used for cutting grass or cereal crops — the smaller scythes were probably a military introduction used to cut hay for horse fodder. The experiments showed that the longer scythes could cut either and it would seem sensible to use it for mowing cereal crops if the tool were available. On the other hand, the site at Farmoor, where one of the scythes was found, produced no evidence of arable agriculture, only of a grassland environment, and certainly within historic times in Britain there is little trace of the consistent use of the scythe for cereal harvesting until the early nineteenth century.

Most surviving Romano-British scythes are of this balanced type but three implements, probably dating to the Roman period, are of a different type. These tools, two from Rushall Down and one from Woodyates, have a different type of attachment device, with tangs which curve round towards the back of the blade. It is difficult to understand how these implements were attached to the handle and how they were used.

Scythes were sharpened in the field by means of hammers and mowers' anvils. These pointed anvils (fig. 30b), with a square domed head for hammering, were set into the ground and were balanced by the two or four coils of iron on either side of the stem of the implement. The scythe blade was put on to the anvil and hammered to sharpen the edge, which would then be given a final polish with a hone. Mowers' anvils are quite common finds on Romano-British sites. Also noteworthy in this context is the fine whetstone found in the hoard which contained the scythes from Silchester.

Other harvesting tools

Pitchforks and rakes were essential tools of the harvest before the industrial revolution and both tools are known from Romano-British sites. Pre-Roman finds of forks are rare, because they would have been made of wood, which has usually not survived. However, one fork dating from the second millennium BC has survived in the waterlogged conditions at a site on the Somerset Levels. It is made of hazel with a slender handle and two tines D-shaped in section. The excavators suggest that the fork, made from a naturally forking piece of wood, was designed for lifting some light substance such as reeds.

Pitchforks of iron (fig. 30 c, d) are fairly common finds from Roman contexts in Britain. They are usually tanged but are occasionally socketed, and a variety of shapes of tines are found. Wooden hay forks must have been commoner, however, as indeed they were until the nineteenth century, and some archaeologists think that the small iron sheaths occasionally found on Roman sites may have been protective tips for the tines of wooden forks. Rakes must also have been usually of wood, though there are frequently found the iron prongs which would have fitted into the wooden head of a rake. Particularly interesting is the surviving head of a rake from Newstead (fig. 30a); the oak head of the rake, 35 centimetres (13¾ inches) long, holds seven curving iron prongs.

Threshing and winnowing

When the grain has been harvested and brought in, the next task is threshing — the extraction of the grain from the husks of the plant. One simple method is to pile the grain on a hardened floor and to lead large animals such as cattle and horses on to it. Gradually the wearing action of their hooves causes the husks to become separated from the grain. This method of threshing is still carried out in some Mediterranean countries on small circular threshing floors. Columella mentions both this method and that of beating the grain with flails. A third method is that of drawing a *tribulum* or threshing sledge, a wooden sledge with flints hammered into the lower surface, on to the grain, so that the cutting action of the flints separates the husks from the grain. Occasionally, flints found on Romano-British sites have been suggested as being *tribulum* flints. Other than these, and the possible identification of a few threshing floors on Romano-British sites, we have no evidence to tell us what was the normal method of threshing in pre-Roman or Roman Britain. Any one of the methods is simple and obvious enough to have been employed.

When the crop is threshed it has to be winnowed so that the separated grain may be collected together for storage or use. If the threshed crop is thrown up into the air with a wooden shovel or fork while a gentle wind is blowing, the wind will blow the lighter chaff away while the heavier grain falls to the ground. This can be done out of doors or inside in barns with two opposite facing doors through which the prevailing wind blows. One wonders if the climate of Britain in the iron age and Roman periods, if not before, made it advisable both to thresh and to winnow indoors if this were possible. The chaff can be further reduced by the use of winnowing baskets and of sieves, which can also be used to select the largest grain for seed. No winnowing forks, shovels, baskets or sieves have been certainly identified so that we have no direct evidence of winnowing techniques, but it is not unreasonable to suppose that these simple and obvious implements were used. The classical agronomists describe the use of all of them in Roman Italy.

The crop

Finally, mention should be made of the crop which was the end product of the soil preparation, sowing, care and harvesting that we have described. The identification of plant remains from archaeological excavations is a science in itself and our understanding of what sort of crops were grown where and at what period in Britain is increasing fast. A primitive form of wheat called emmer *(Triticum dicoccum)* seems to have been the main wheat grown in early periods in Britain; later introductions from the bronze age onwards were spelt *(Triticum spelta)*, bread wheat *(Triticum aestivum)* and club wheat

(Triticum compactum). The last two are more cultivated wheats, known as naked wheats because their grains become loose within their casing at maturity and hence are more easily threshed than the uncultivated hulled wheats. More useful in the marginal areas of Britain than the less tolerant wheat was a barley known as *Hordeum tetrastichum,* grown in Britain from the neolithic period onwards.

How these crops were stored when they had been threshed remains a controversial issue among archaeologists, and storage is likely to have been done in different ways in different areas of Britain. Underground storage pits, often lined with material such as basketry, may have been one method of grain storage in the south of Britain during the iron age, while the four-post structures, long ago interpreted as granaries, a common feature in iron age sites in England at least, have not been discounted as grain storage structures. Storage methods previous to the iron age are unknown.

Beans and peas were cultivated probably from the neolithic and bronze age onwards, as the plant remains from Ogmore and Grimes Graves show. The presence of grape pips in Roman contexts on some excavated sites shows that the vine may have been grown in Britain at this time, but the absence in Britain of the distinctive viticulture knives, common in Roman Gaul for example, suggests that the Romans considered Britain to be as inappropriate a country for wine producing as continental competitors would have us believe it remains to this day.

List of classical agronomists

Cato. *De Agricultura.* Mid second century BC.
Columella. *De Re Rustica* and *De Arboribus.* First century AD.
Palladius. *Opus Agriculturae.* Late fourth century AD.
Pliny. *Natural History.* First century AD.
Varro. *De Re Rustica* and *De Lingua Latina.* Mid first century BC.
Vergil. *Georgics.* First century BC.

Further reading

Applebaum, S. 'Roman Britain (1. AD 43 to 1043)'. Volume 1: ii in Finberg, H.P.R. (ed) *The Agrarian History of England and Wales.* 1972.

Bowen, H. C. *Ancient Fields.* British Association for the Advancement of Science. 1961.

Bowen, H. C. and Fowler, P. J. *Early Land Allotment: A Survey of Recent Work.* British Archaeological Reports no. 48. 1978.

Coles, J. M. *Archaeology by Experiment.* 1973.

Curwen, E. C. *Plough and Pasture.* 1948.

Evans, J. G. *The Environment of Early Man in the British Isles.* 1975.

Fenton, A. 'Early and Traditional Cultivating Implements in Scotland'. *Proceedings of the Society of Antiquaries of Scotland.* XCVI (1962-3), 264-317.

Fox, C. 'The Socketed Bronze Sickles of the British Isles'. *Proceedings of the Prehistoric Society* 5 (1939), 222-48.

Fox, C. 'The Non-Socketed Bronze Sickles of Britain'. *Archaeologia Cambrensis* XCVI (1941), 136-62.

Gailey, A. and Fenton, A. *The Spade in Northern and Atlantic Europe.* Belfast, 1970.

Glob, P. V. *Ard og Plov i Nordens Oltid* (with English summary). Aarhus, 1951.

Manning, W. H. 'The Plough in Roman Britain'. *Journal of Roman Studies* 54 (1964), 54-65.

Manning, W. H. 'The Piercebridge Ploughgroup'. *Prehistoric and Roman Studies.* British Museum (1971), 125-36.

Rees, S. E. *Agricultural Implements in Prehistoric and Roman Britain.* British Archaeological Reports no. 69. 1979.

Reynolds, P. *Iron-Age Farm: The Butser Experiment.* London, 1979.

Steensberg, A. *Ancient Harvesting Implements.* Copenhagen, 1943.

Taylor, C. *Fields in the English Landscape.* London, 1975.

White, K. D. *Agricultural Implements of the Roman World.* Cambridge, 1967.

White, K. D. *Roman Farming.* London, 1970.

White, K. D. *Farm Equipment of the Roman World.* Cambridge, 1975.

See also volumes of the journal *Tools and Tillage,* a journal published in Copenhagen and dealing exclusively with the history of agricultural tools.

Museums

Many museums have interesting displays devoted to the agricultural practices of prehistoric and Romano-British people in the area. The museums listed below have extensive collections of agricultural tools or smaller but particularly interesting groups of tools.

Cambridge: University Museum of Archaeology and Ethnology, Downing Street. Important collections of tools from Great Chesterford, including the long scythes and turf cutter. Spade shoe and pruning knives from Worlington.

Cardiff: National Museum of Wales, Cathays Park. Llyn Fawr hoard with bronze and iron sickles; Llyn Cerrig Bach hoard with balanced sickle. Iron winged share from Dinorben; wooden shares from Walesland Rath.

Devizes: Devizes Museum, Long Street. Collections of reaping hooks and pruning hooks from iron age and Roman sites; the iron share from Box.

Edinburgh: National Museum of Antiquities of Scotland, Queen Street. Four iron scythes, hoes, and the rake head from Newstead. Stone ard points from Northern Isles.

Glastonbury: Glastonbury Lake Village Museum, The Tribunal, High Street. Rich collection of billhooks, sickles, hooks, parts of spades from iron age sites at Glastonbury and Meare.

London: Museum of London, London Wall, EC2. Flanged symmetrical iron shares, spade shoes, sickles and pruning hooks from sites in London.

Manchester: Manchester Museum, The University. Much of the first-century ironwork from Bigbury including the large slashing tools.

Northampton: Northampton Central Museum, Guildhall Road. Iron age ploughshares from Hunsbury, iron scythe from Irchester.

Peterborough: Peterborough Museum and Art Gallery, Priestgate. Roman hoard from Sibson containing iron scythe and coulter.

Reading: Reading Museum and Art Gallery, Blagrave Street. Important collection of Roman ironwork from Silchester including bar shares, coulters, mowers' anvils.

St Albans: Verulamium Museum, St Michaels. A collection of spade shoes from Verulamium.

Woodstock: Oxfordshire County Museum, Fletcher's House. Reaping hooks from iron age and Roman sites, wooden ard share from Abingdon, long scythes from Farmoor and Hardwick.

In addition to the above, the folk museums at Reading (Museum of English Rural Life, Reading University) and Cardiff (Welsh Folk Museum, St Fagans) have large collections of pre-industrial revolution agricultural tools which are most useful for illustrating how earlier tools were used. The experimental iron age farm at Butser Hill, Hampshire, has displays of iron age type crops and has occasional demonstrations of how early farming tools were used.

Plate 1. Wooden ard share from Walesland Rath. (Copyright: National Museum of Wales.)
Plate 2. Stone ard tip from Orkney.

Plate 3. Wear marks on upper surface of ard tip.

Plate 4. Wear marks on side of ard tip.

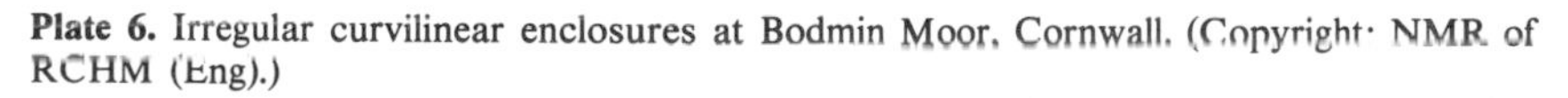

Plate 5. 'Celtic' fields at Smacam Down, Dorset. (Copyright NMR of RCHM (Eng).)

Plate 6. Irregular curvilinear enclosures at Bodmin Moor, Cornwall. (Copyright: NMR of RCHM (Eng).)

Plate 7. Criss-cross ard traces at Skaill, Orkney. (Photograph: P. S. Gelling.)

Plate 8. Cross-section through ard trace, showing asymmetric V-shaped profile.

Plate 9. Reconstruction of Donneruplund ard used in ploughing experiments in 1956/7. (Copyright: NMR of RCHM (Eng).)

Plate 10. The reconstructed ard in use. (Copyright: NMR of RCHM (Eng).)

Plate 11. Reconstruction of a yoke used in ploughing experiments. (Copyright: NMR of RCHM (Eng).)

Plate 12. The reconstructed ard and plough team. (Copyright: NMR of RCHM (Eng).)

Plate 13. *(Left)* The two bronze sickles and one iron sickle from the Llyn Fawr hoard, Glamorgan. (Copyright: National Museum of Wales.)
Plate 14. *(Right)* Iron balanced sickle from the hoard from Llyn Cerrig Bach, Anglesey. (Copyright: National Museum of Wales.)

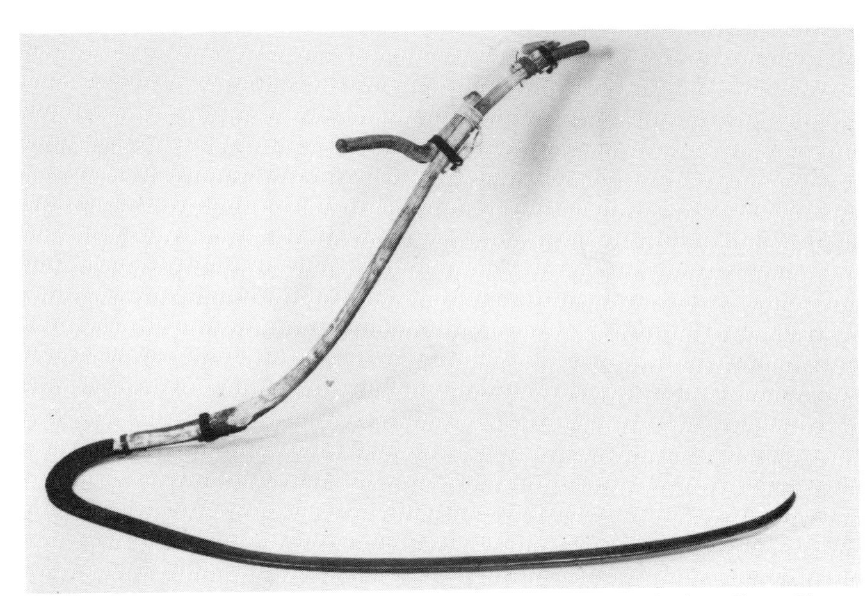

Plates 15 and 16. Reconstructed haftings of a copy of a Roman scythe from Great Chesterford, used in experimental mowing. (Copyright: Museum of English Rural Life, Reading.)

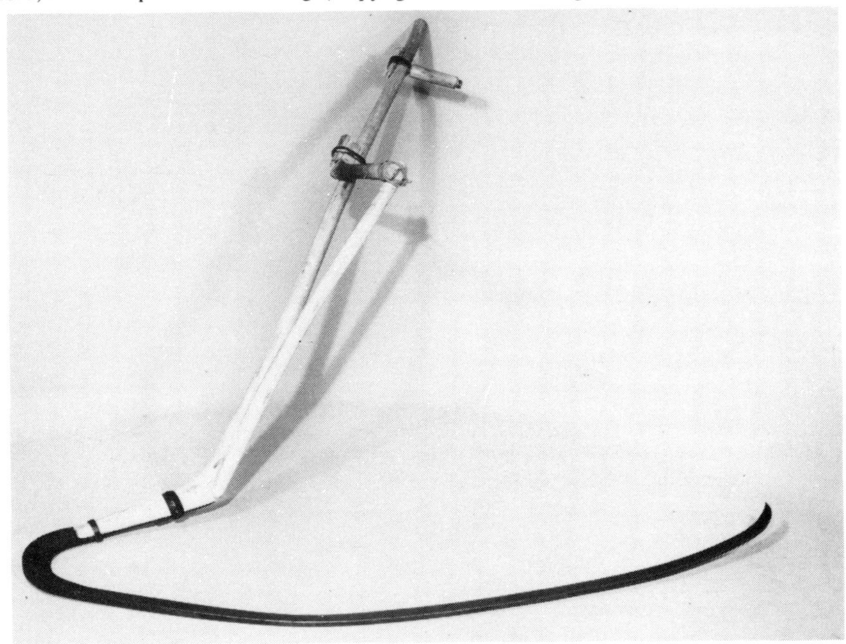

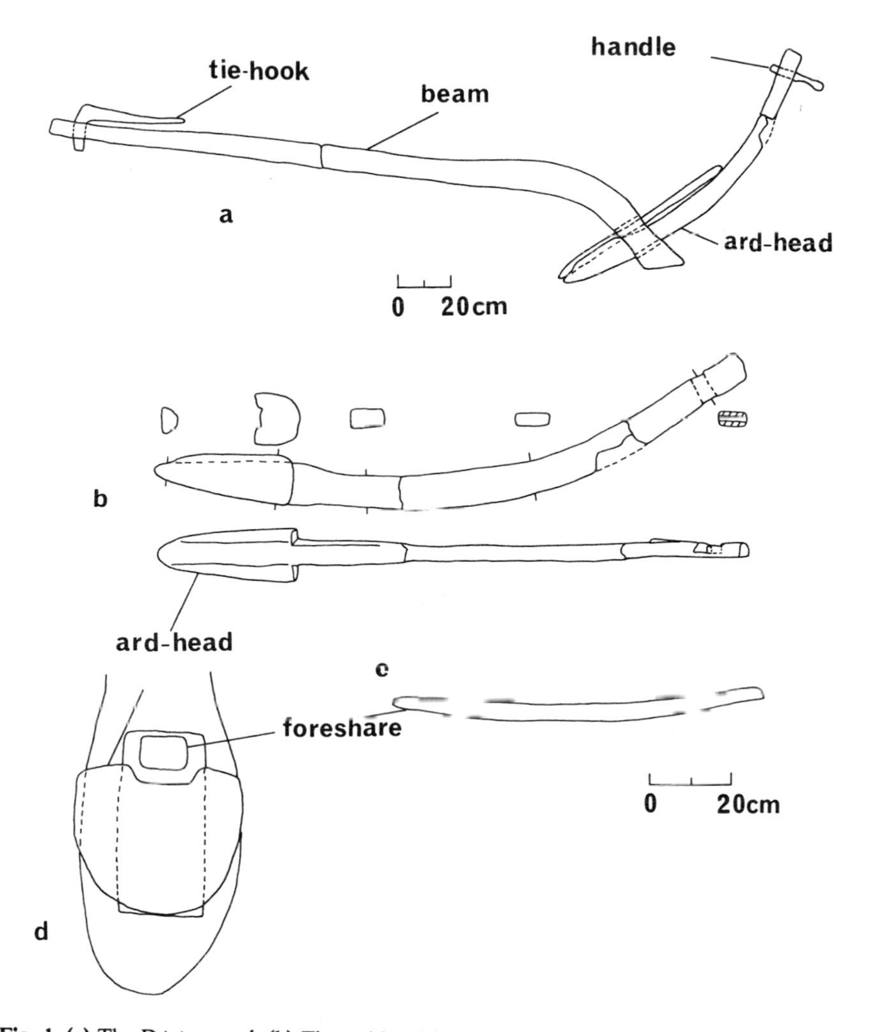

Fig. 1. (a) The Døstrup ard. **(b)** The ard-head from the Døstrup ard: side view (upper), top view (lower). **(c)** The foreshare from the Døstrup ard. **(d)** The cross-section through the mortise in the beam of the Døstrup ard showing positions of the foreshare and ard-head in the mortise.

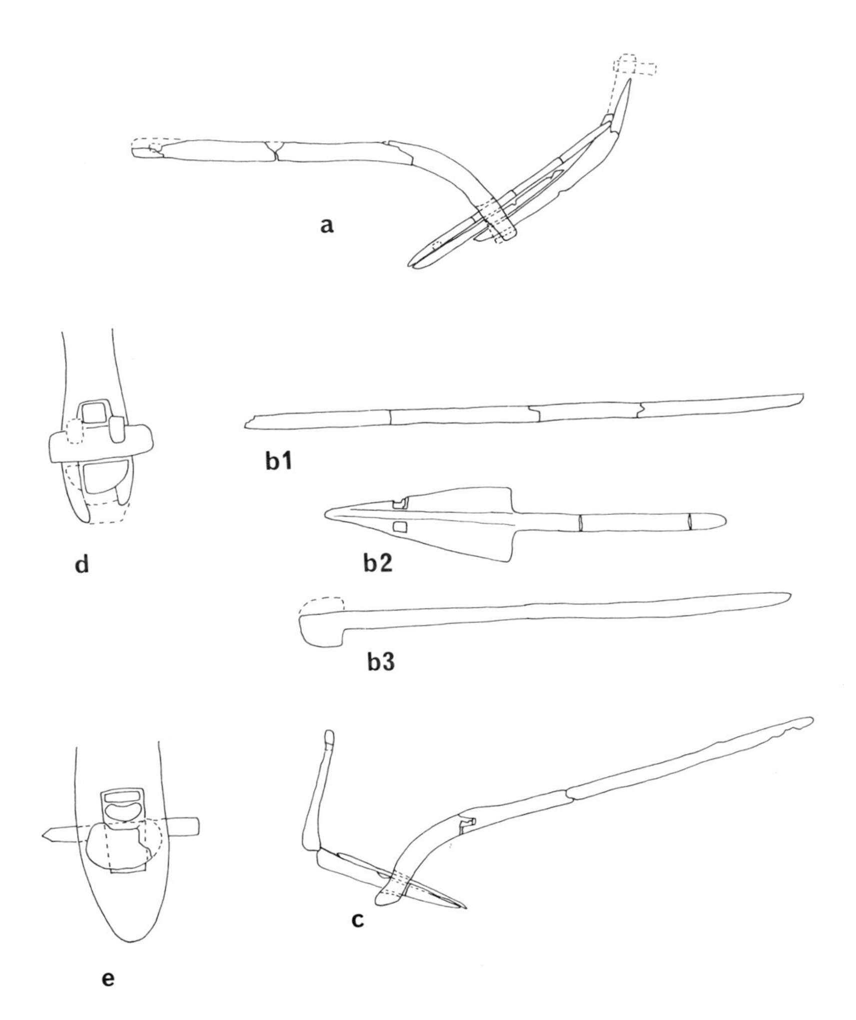

Fig. 2. (a) The Donneruplund ard. **(b)** The foreshare (1), arrow-shaped main share (2) and ard-head (3) of the Donneruplund ard. **(c)** The Hendriksmose ard. **(d, e)** Cross-sections through the beam mortise of the Donneruplund ard (d) and the Hendriksmose ard (e) with hypothetical wedge reconstructions in place, as used in experiments.

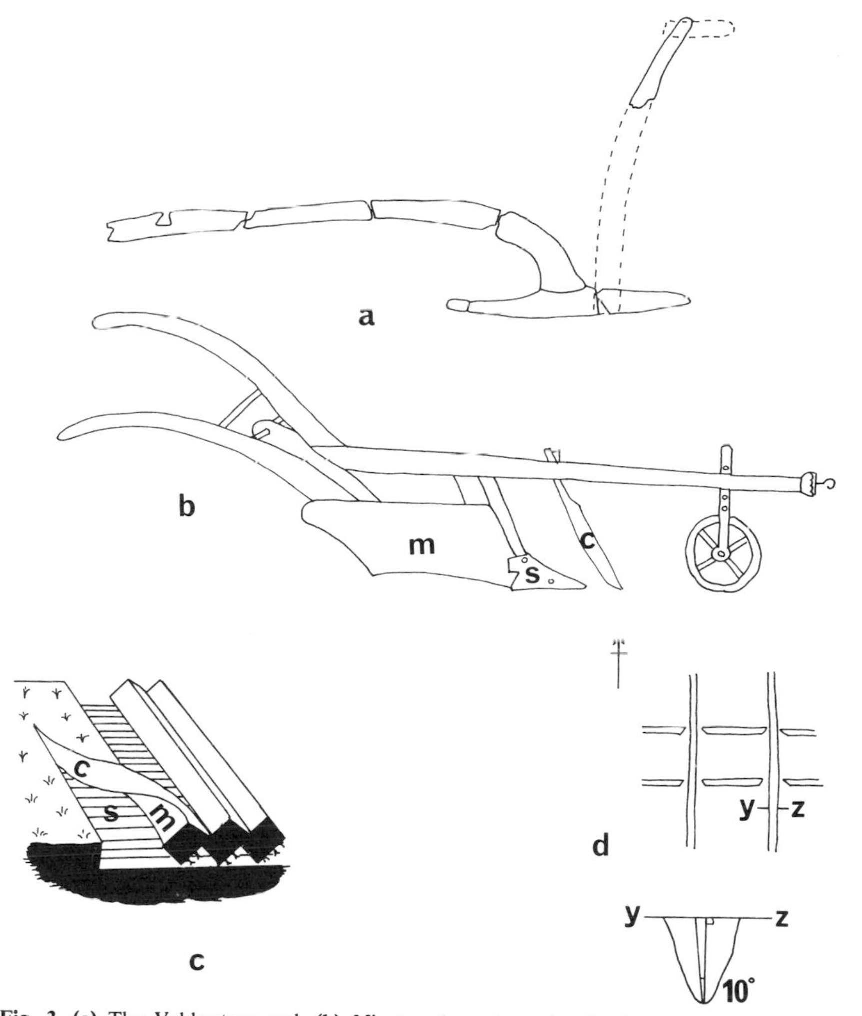

Fig. 3. (a) The Vebbestrup ard. **(b)** Nineteenth-century plough showing positions of the coulter (C), share (S) and mouldboard (M). **(c)** Diagram to show the functions of the coulter (C), which cuts the sod vertically, the share (S), which undercuts the sod horizontally, and the mouldboard (M), which turns the cut sod (after Bowen). **(d)** Diagram to show grid pattern of criss-cross ard marks. The two north-south marks have cut through the earlier east-west marks leaving an arrow-shaped mark through the earlier marks, indicating that the later marks resulted from the ard being dragged northwards. The cross-section of the mark at Y-Z shows an asymmetric V-profile. The angle between the line bisecting the V and the line running at 90 degrees to the ground surface is called the angle of tilt. As the ard was running northwards, it follows that at Y-Z the ard was being tilted at 10 degrees to the left-hand side.

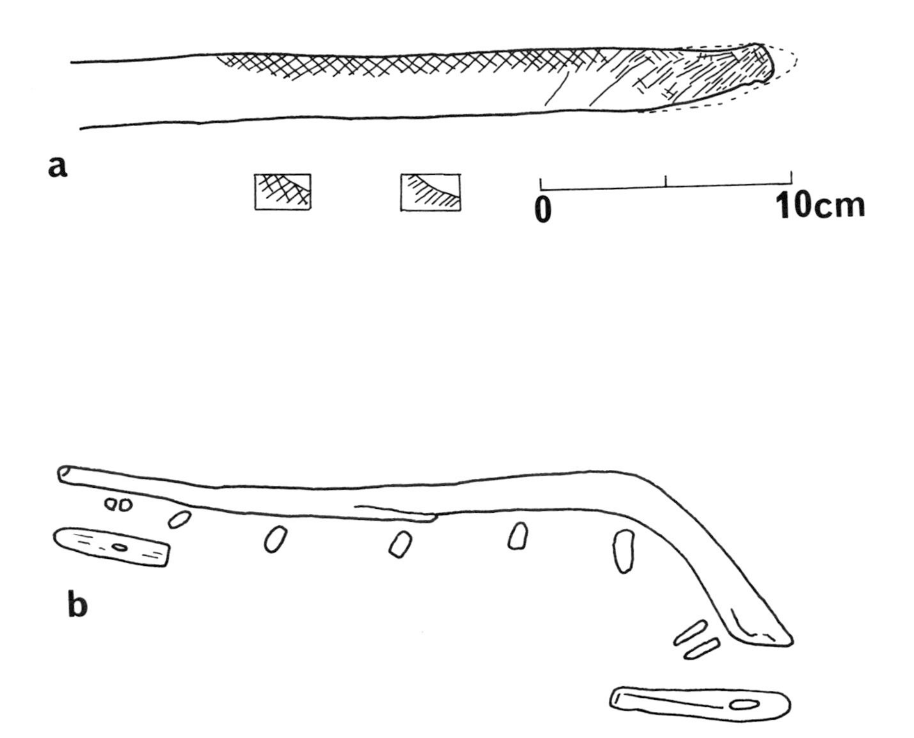

Fig. 4. (a) The wear pattern displayed by the wooden foreshare of the reconstruction based on the Hendriksmose ard after experimental ploughing (after Hansen). **(b)** The Lochmaben ard beam.

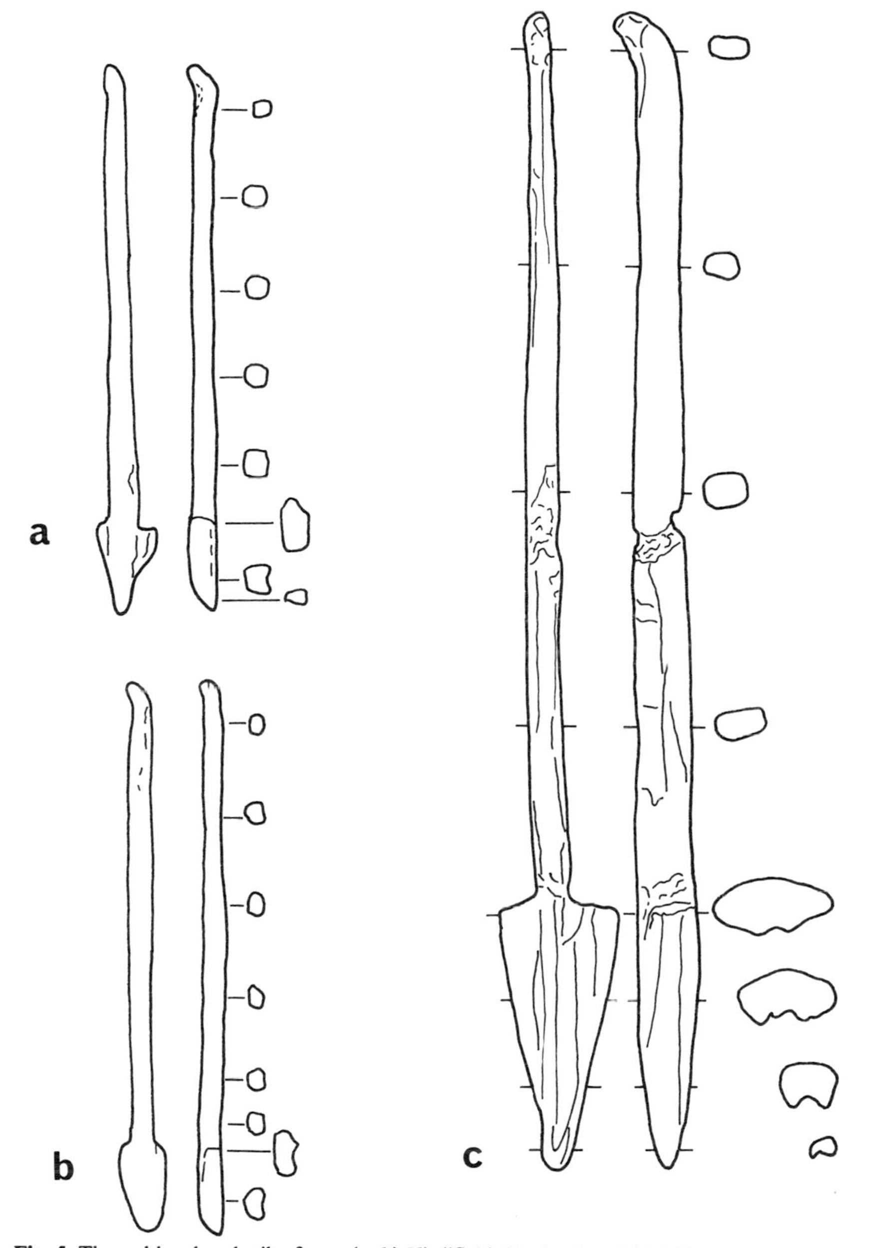

Fig. 5. The ard-head and stilts from: **(a, b)** Virdifield, Shetland; and **(c)** Milton Loch crannog. Not to scale.

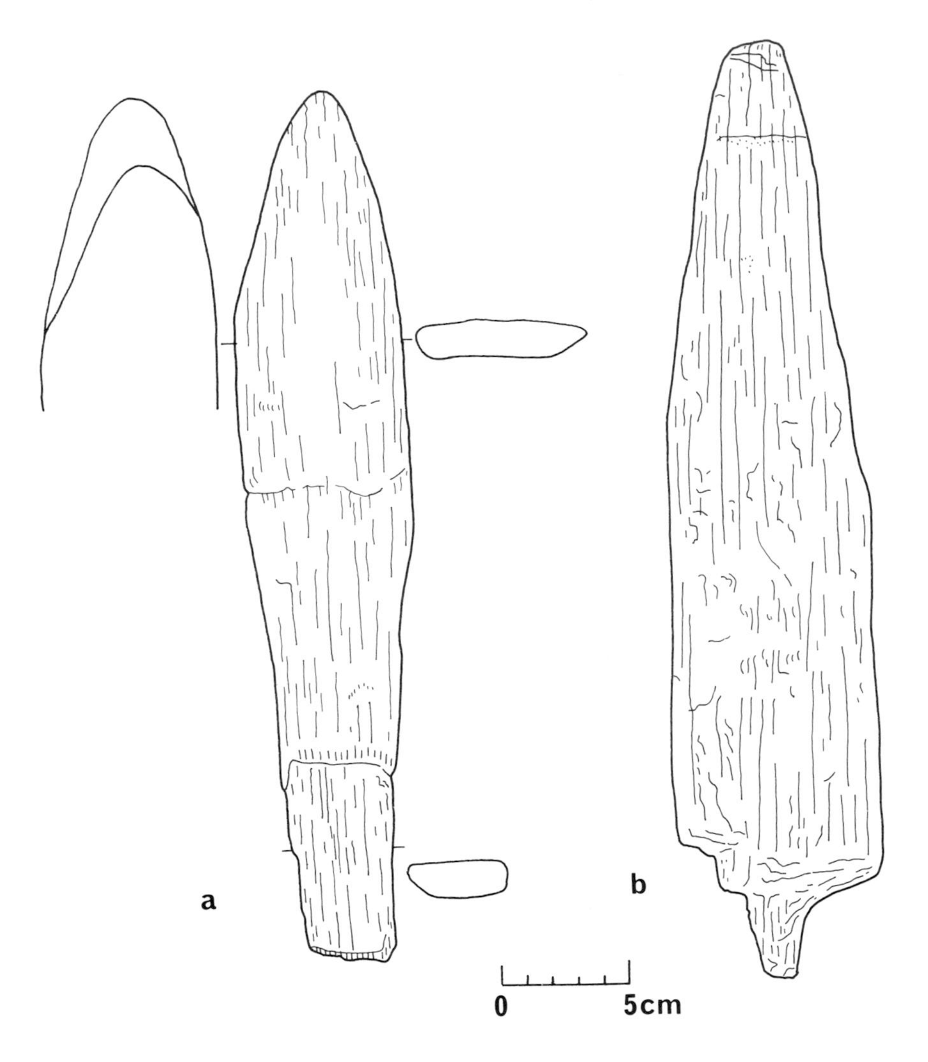

Fig. 6. The wooden ard shares from **(a)** Walesland Rath and **(b)** Abingdon.

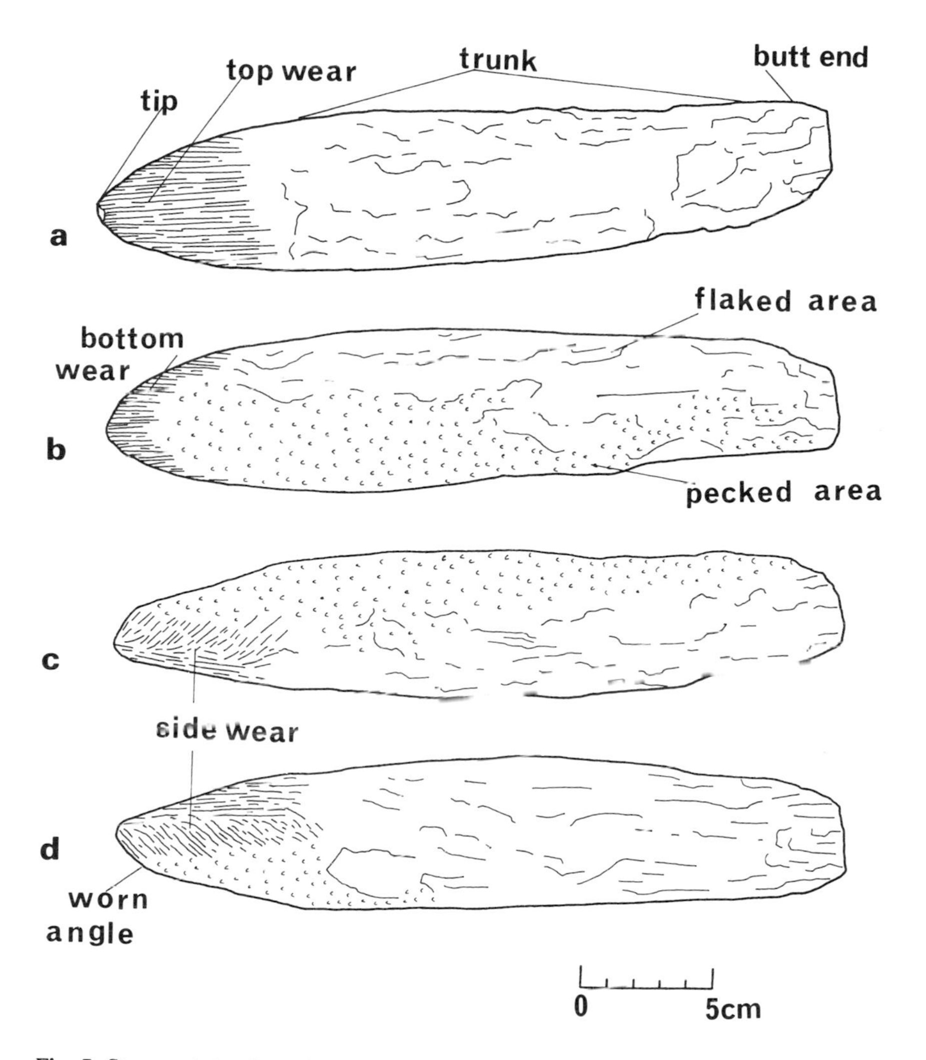

Fig. 7. Stone ard tips from Shetland. **(a)** Upper surface. **(b)** Lower surface. **(c, d)** Sides.

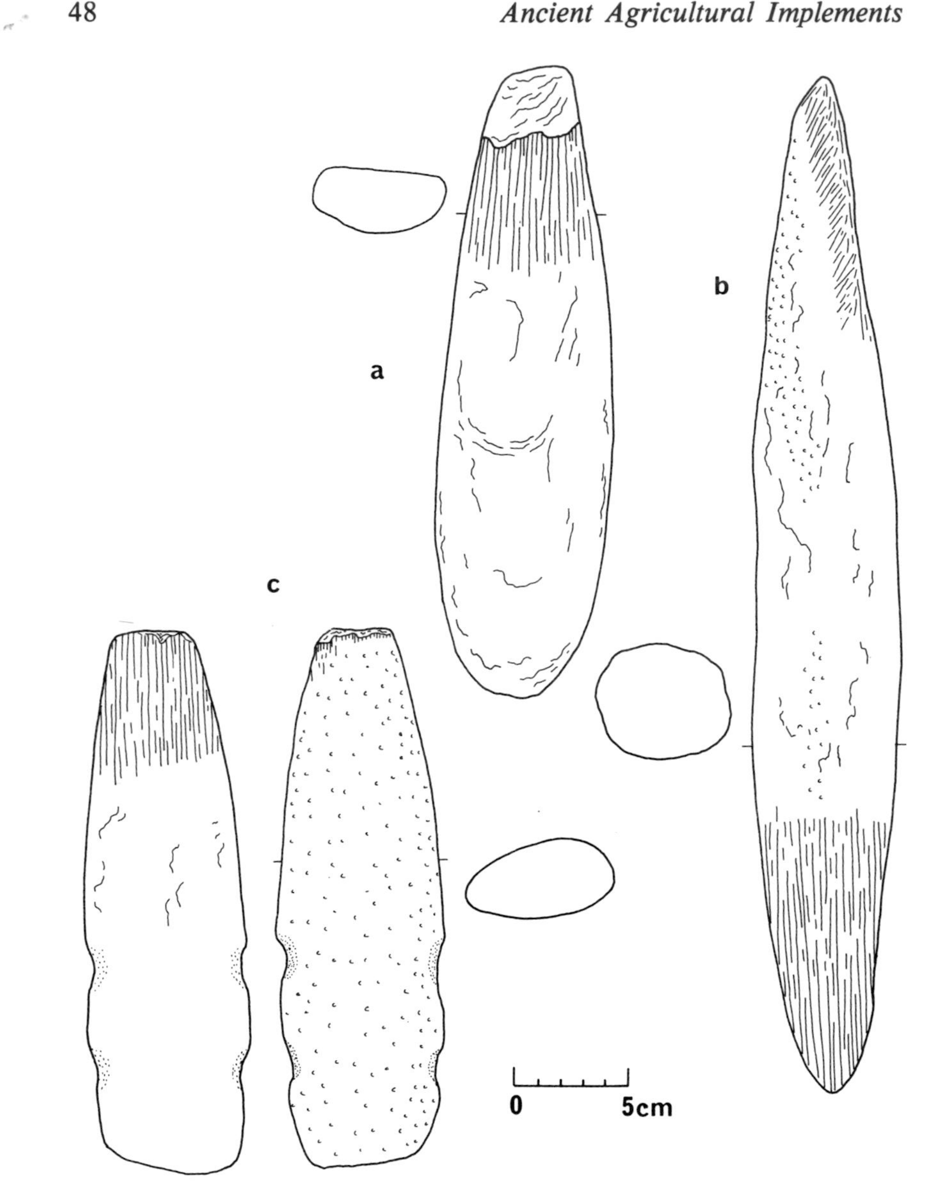

Fig. 8. Stone ard tips: **(a)** with tapered butt end, **(b)** double-pointed, worn at both ends, and **(c)** with squared butt end, with side grooves.

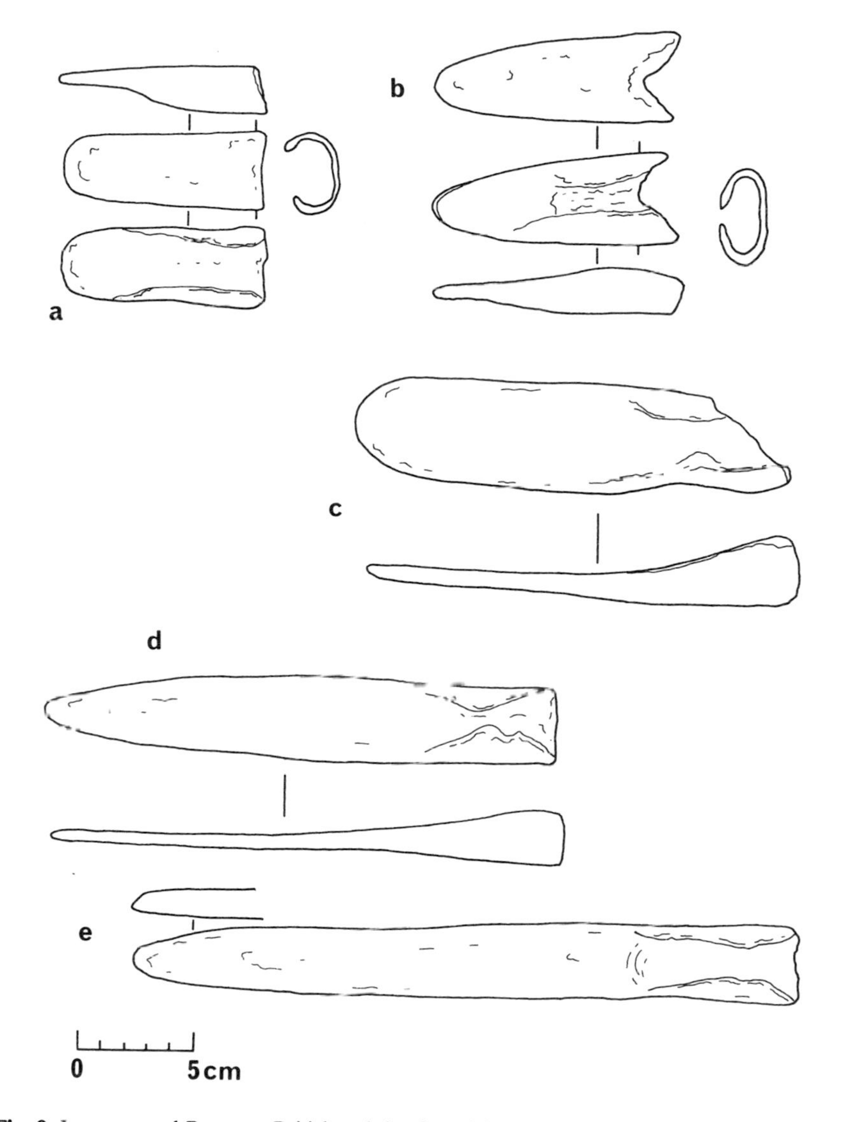

Fig. 9. Iron age and Romano-British ard tips from **(a)** Crayford, **(b)** Hunsbury, **(c)** Frilford, **(d)** Walthamstow and **(e)** Bloxham.

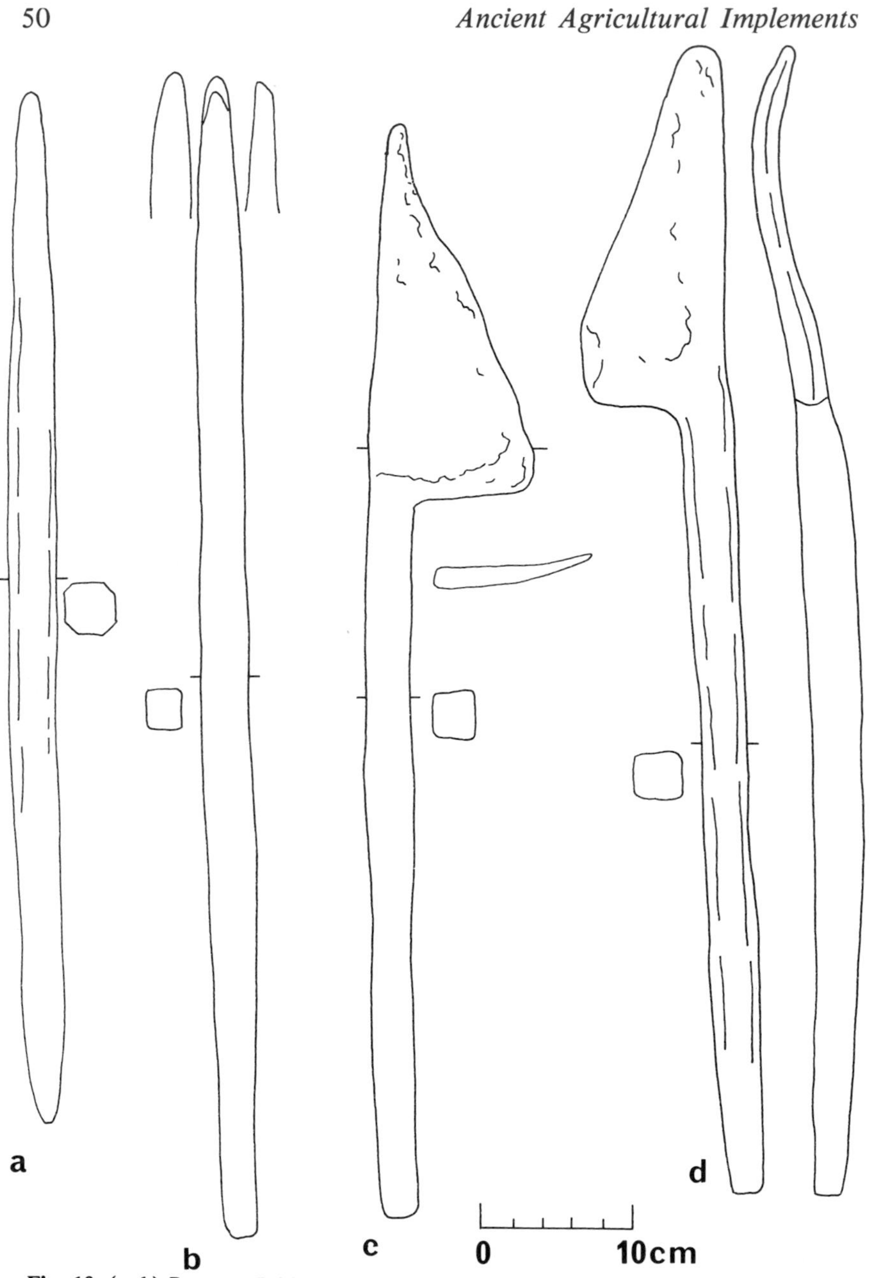

Fig. 10. (a, b) Romano-British iron bar shares from Silchester. **(c, d)** Romano-British iron coulters from Great Whitcombe and Stanton Low.

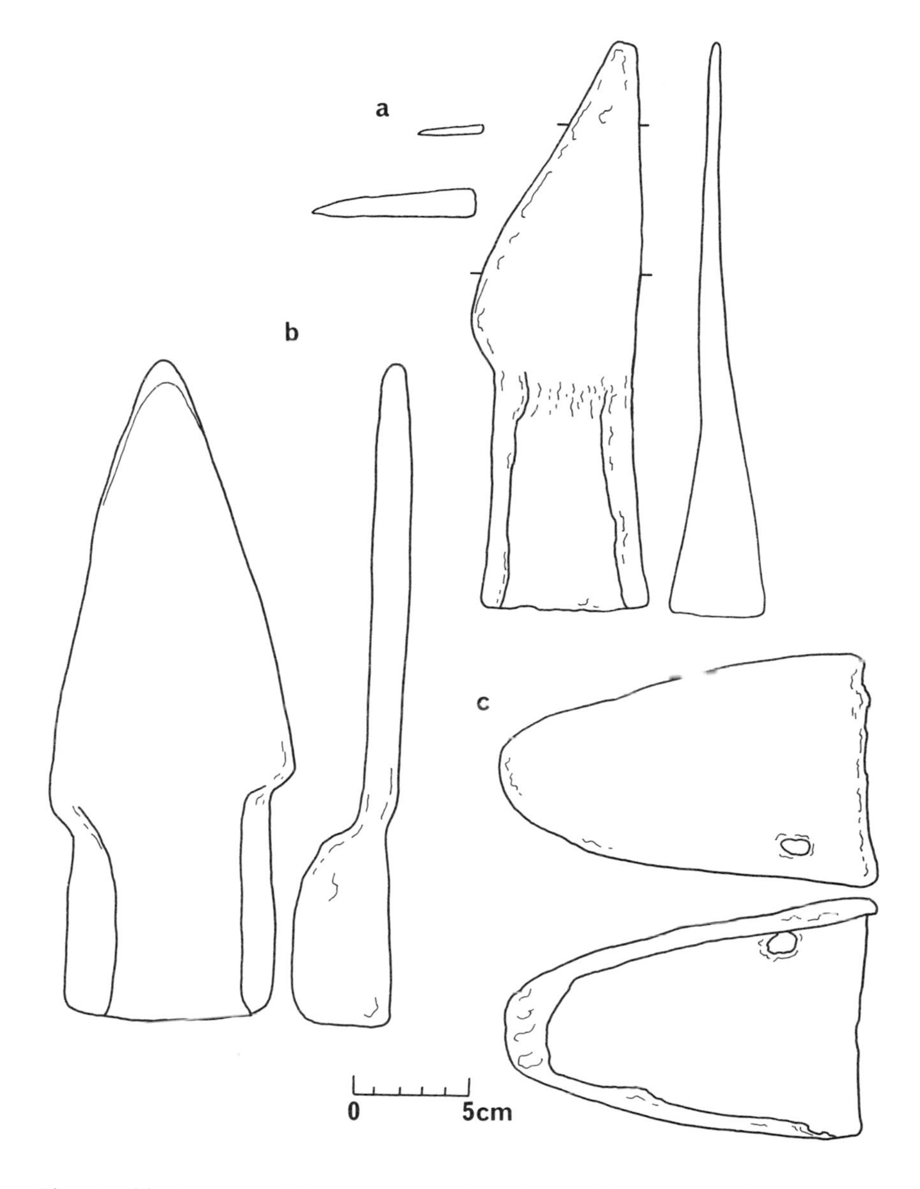

Fig. 11. (a) Romano-British winged iron share from Folkestone. **(b)** Romano-British symmetrical flanged share from London. **(c)** Romano-British share from Blackburn Mill.

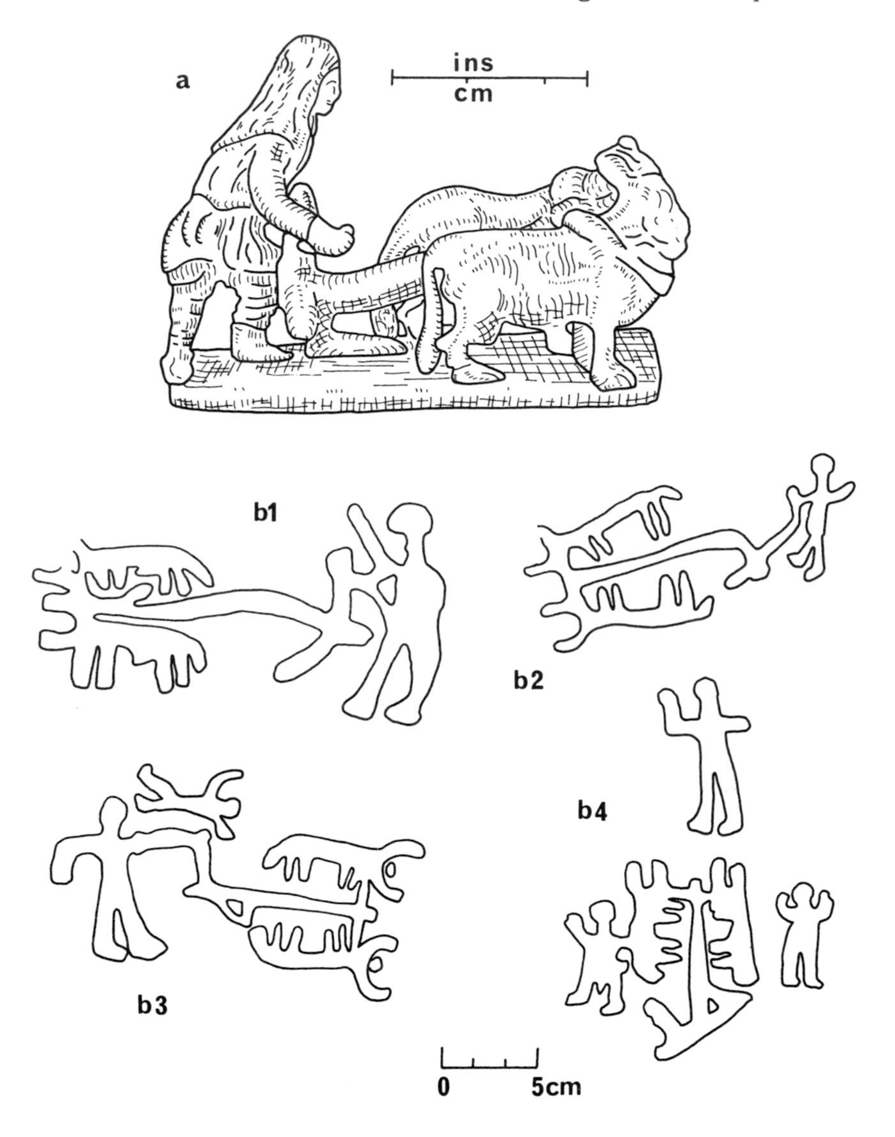

Fig. 12. (a) The Romano-British bronze model of the Piercebridge ploughgroup drawn without casting strip (after Manning). **(b)** Scandinavian rock engravings of ploughing scenes probably dating to the bronze age from (1) Aspeberg, (2) Valla Östergård and (3) Finntorp (with bow ard), (4) Finntorp (with crook ard) (after Glob).

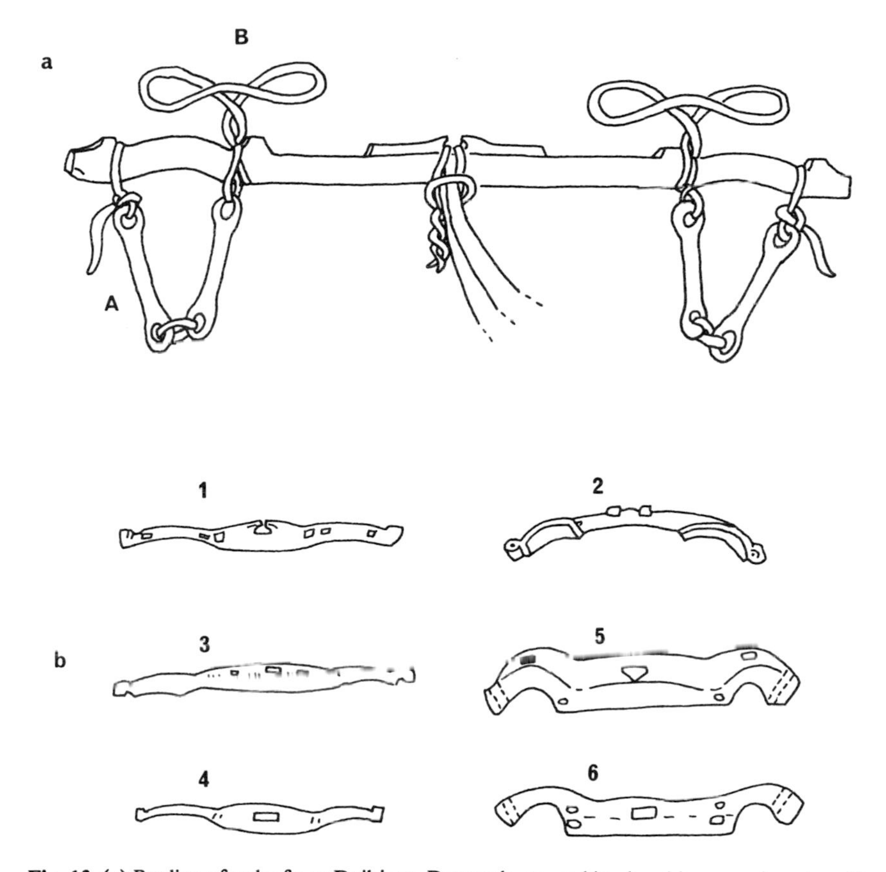

Fig. 13. (a) Replica of yoke from Dejbjerg, Denmark, as used in ploughing experiments, with neck ties (A) and noose fastenings (B) for the horns of the oxen at the top (after Hansen). **(b)** Horn or head yokes from Scotland and Ireland (1, 2, 3, 4). Withers yokes from Ireland (5, 6). Not to scale (after Fenton).

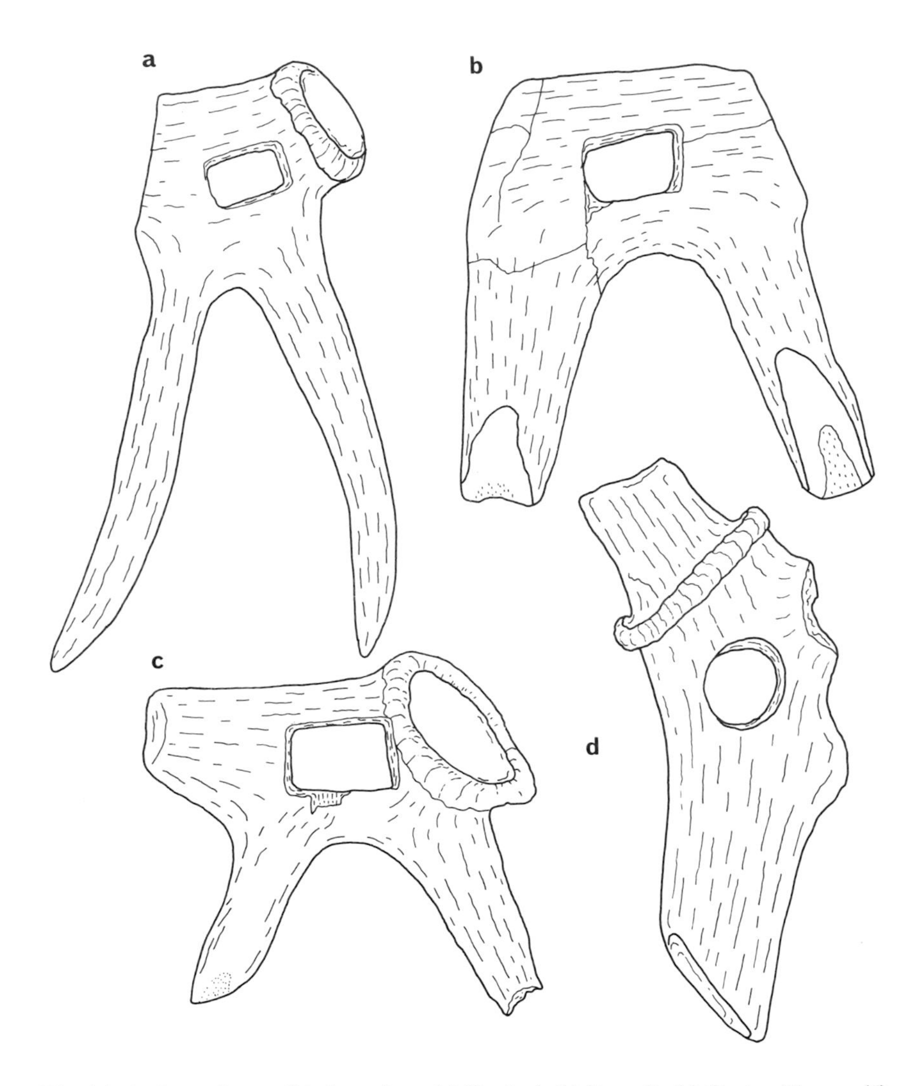

Fig. 14. Antler tools, possibly hoes, from **(a)** Sleaford, **(b)** Darenth, **(c)** Gayton Thorpe, **(d)** Feltwell. All Romano-British except (d) which is probably bronze age. All about $\frac{2}{3}$ size.

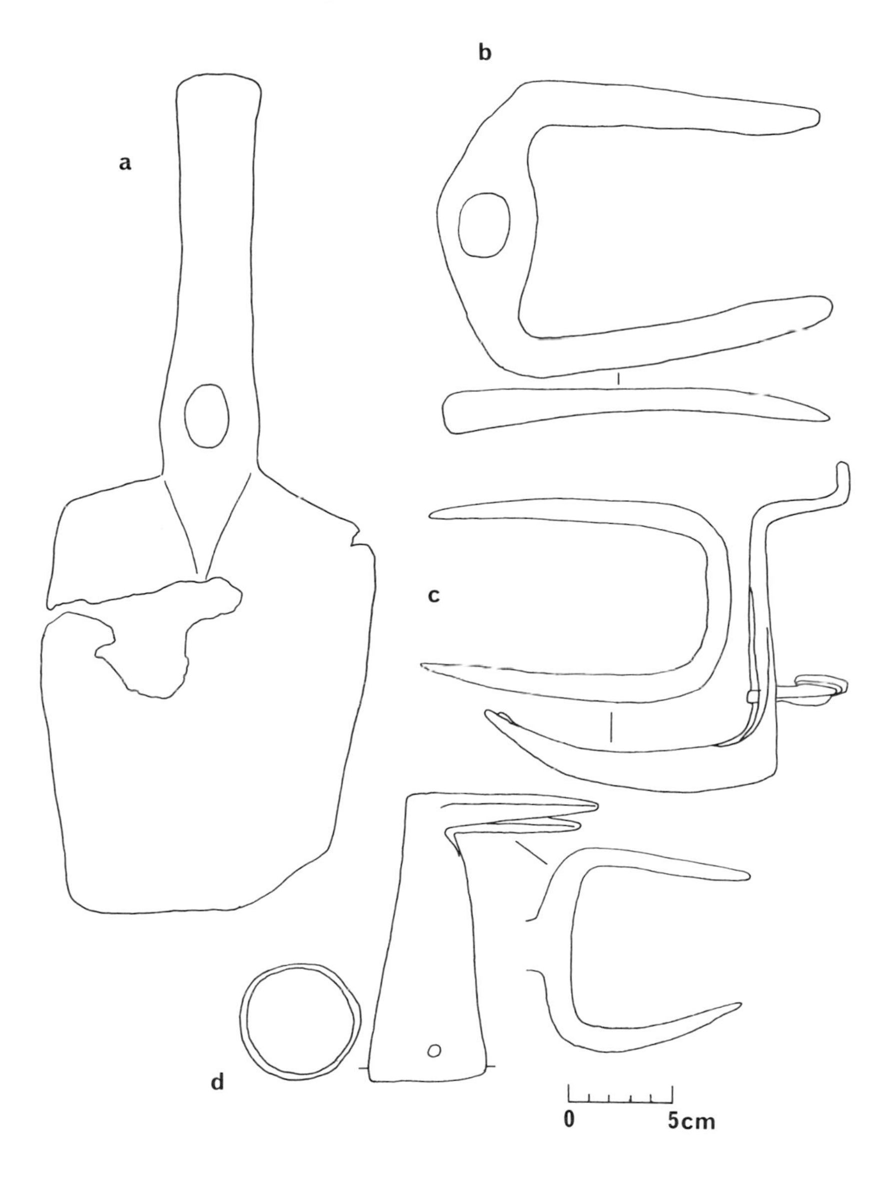

Fig. 15. (a) Romano-British entrenching tool from London. **(b-d)** Romano-British double pronged hoes from Coygan Camp, Rushall Down and London.

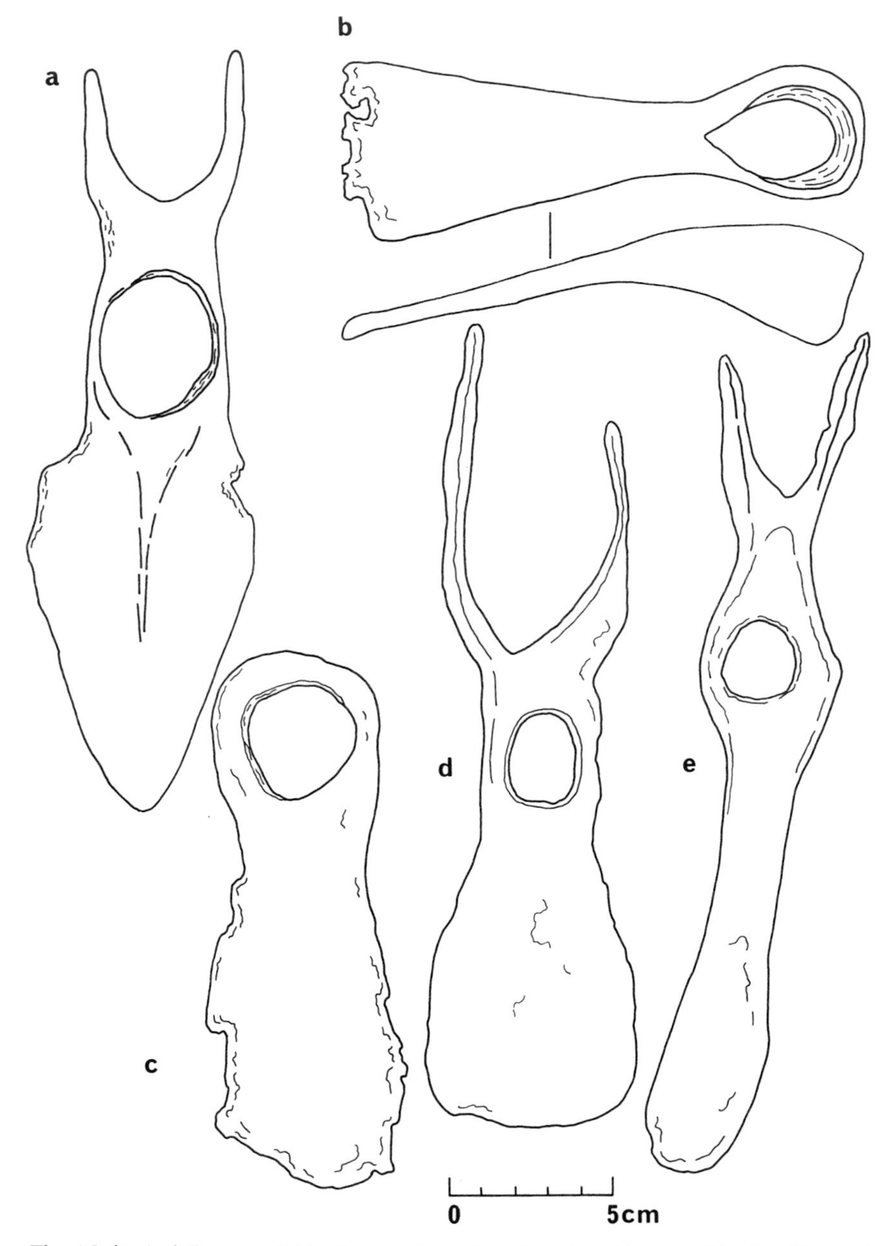

Fig. 16. (a, d, e) Romano-British iron *ascia-rastrum* hoes from London, Thealby Mine and Silchester. **(b, c)** Romano-British single-bladed hoes from Loudon Hill and Nottingham.

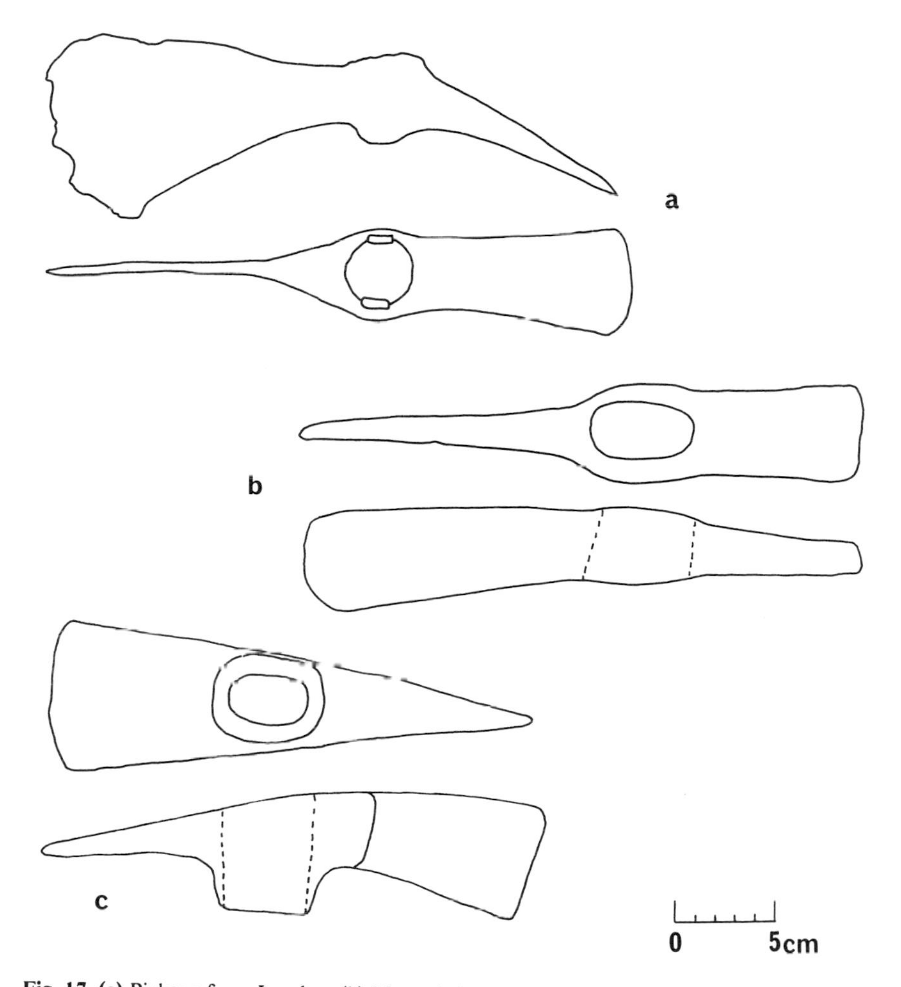

Fig. 17. (a) Pickaxe from London. **(b)** Mattock from Bigbury. **(c)** Mattock from Lakenheath.

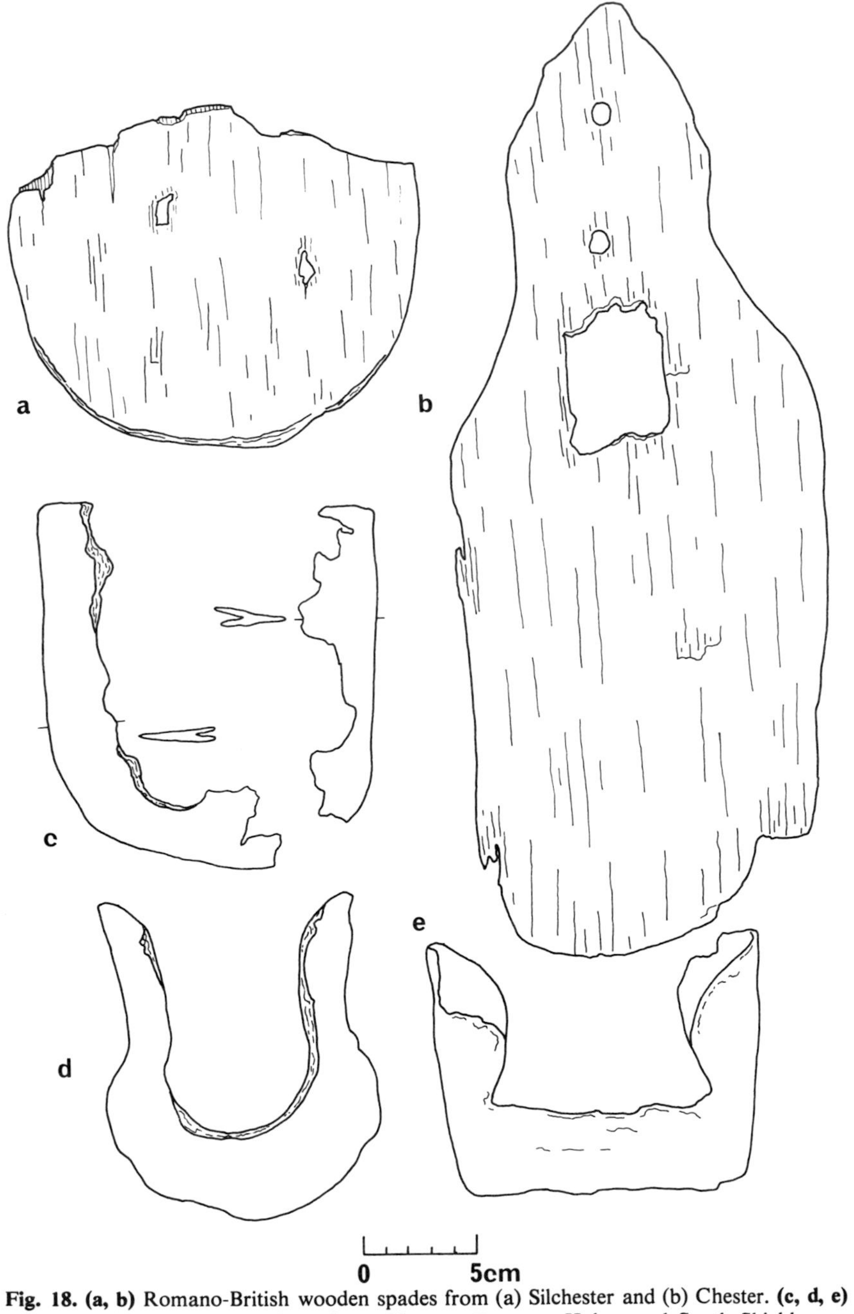

Fig. 18. (a, b) Romano-British wooden spades from (a) Silchester and (b) Chester. **(c, d, e)** Romano-British iron spade shoes from Chesters, Runcton Holme and South Shields.

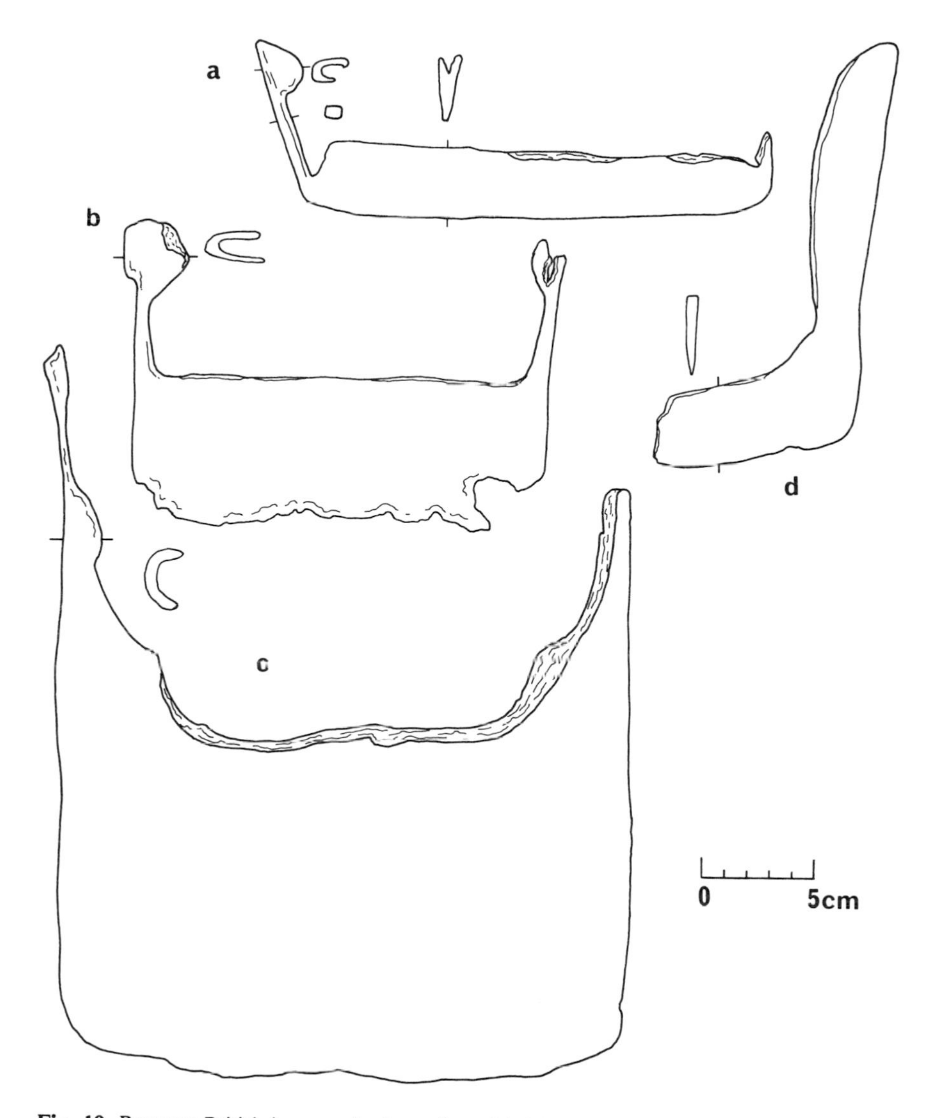

Fig. 19. Romano-British iron spade shoes from **(a)** Silchester, **(b)** Spoonley, **(c)** Colchester and **(d)** Caistor.

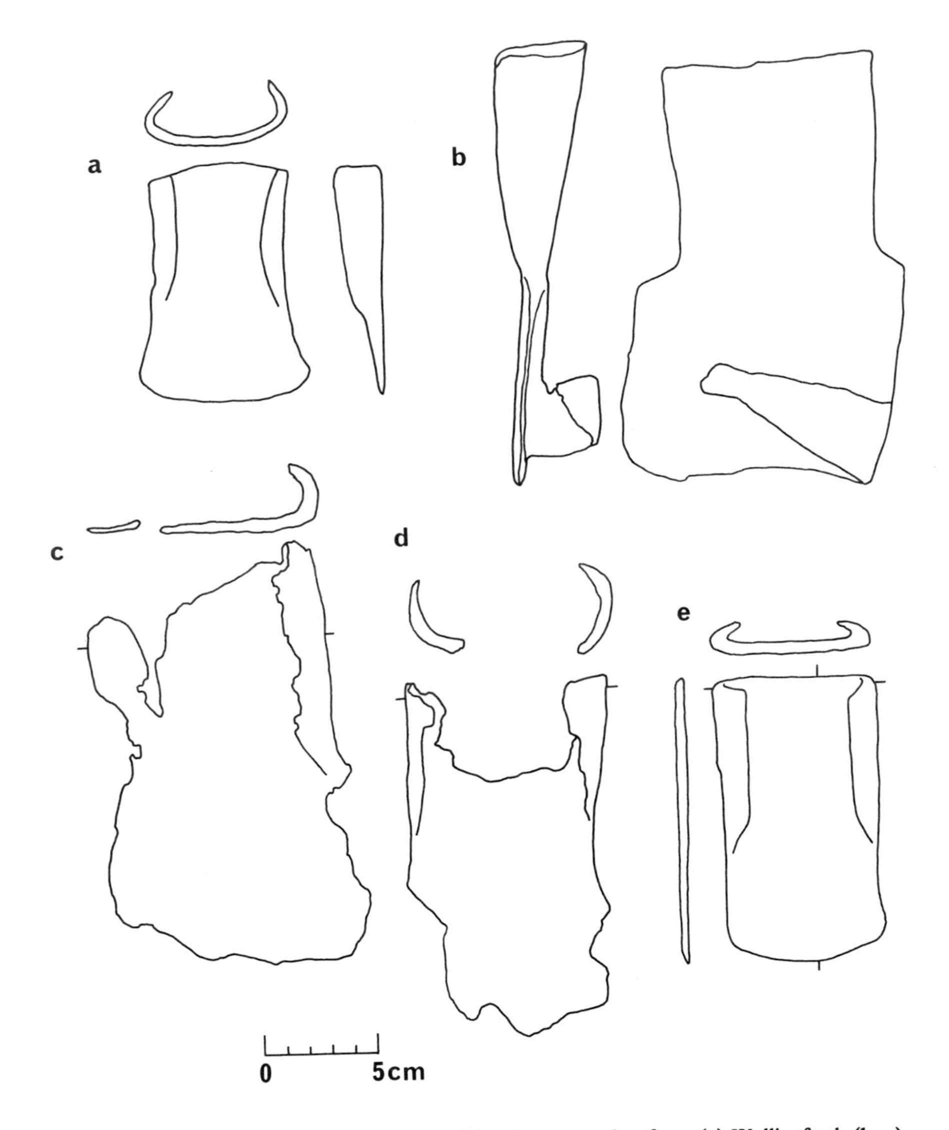

Fig. 20. Romano-British iron blades, probably of peat spades, from **(a)** Wallingford, **(b, c)** Blackburn Mill, **(d)** Corbridge and **(e)** Silchester.

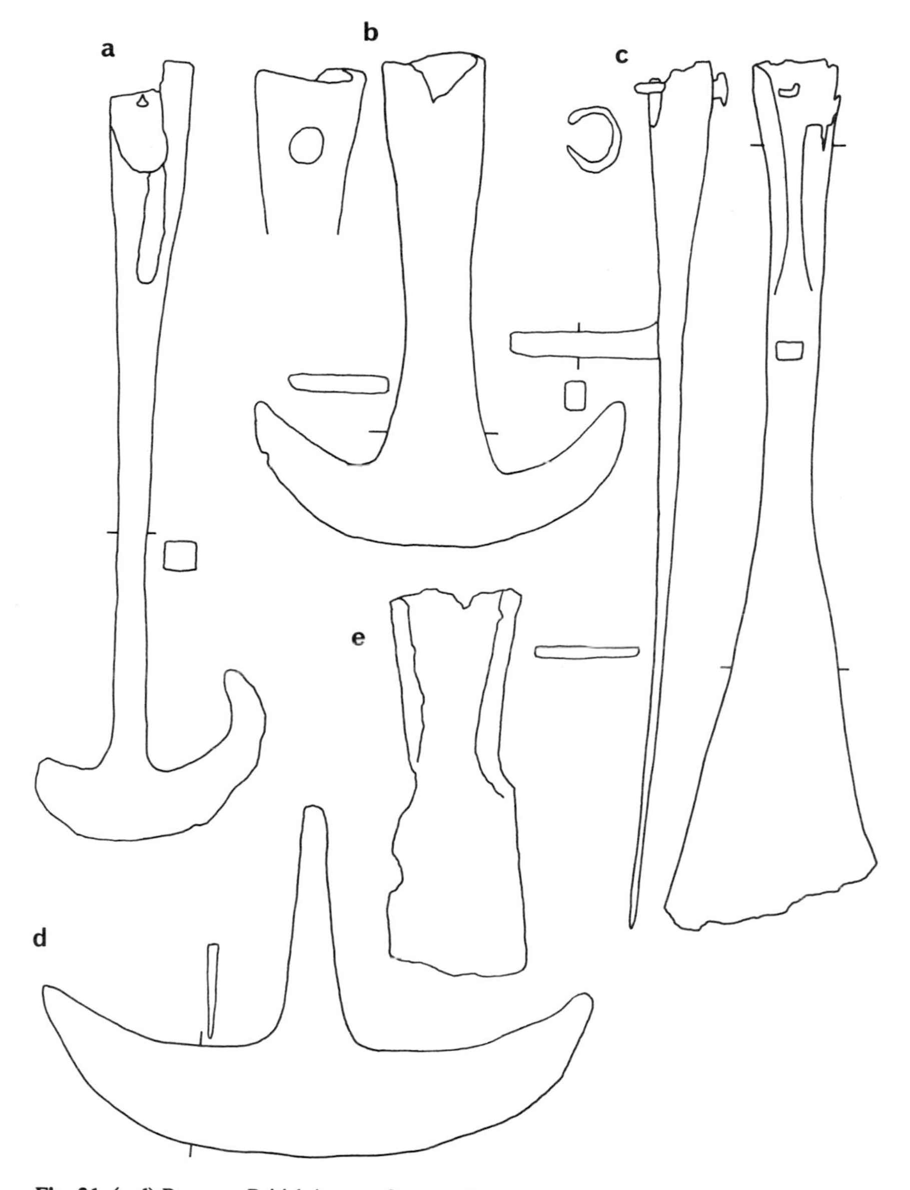

Fig. 21. (a-d) Romano-British iron turf cutters from Housesteads, Newstead, Great Chesterford and South Shields. **(e)** Romano-British iron spud from London. All approximately3/5 size.

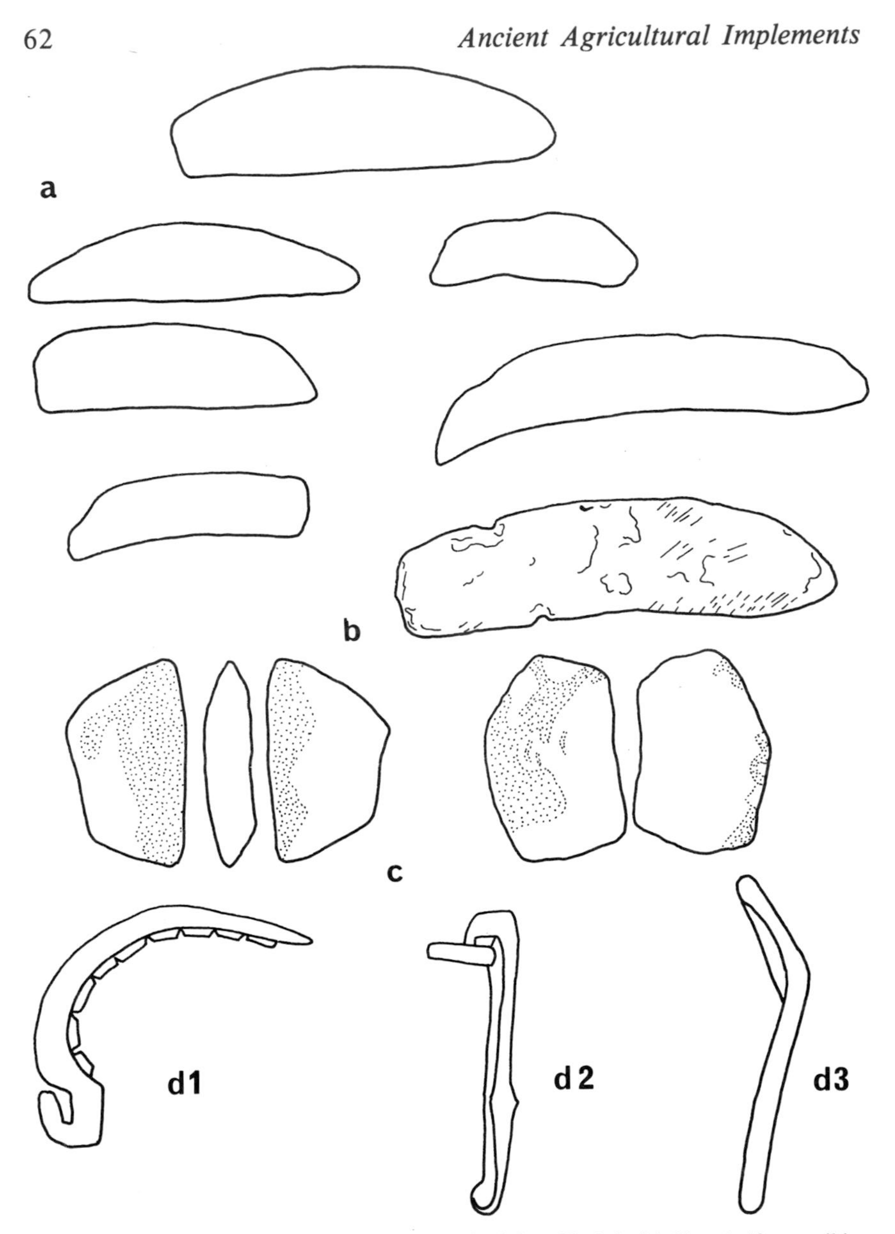

Fig. 22. (a) Crescentic flint sickles from Britain (after Clarke). **(b)** Slate knife, possibly a sickle, from Jarlshof, Shetland. **(c)** 'Sickle-flints' (after Curwen). Stippling shows areas of gloss. **(d)** (1) Flint toothed sickle from the tomb of Hemaka, Egypt (after Childe). (2) Flint blade in original wooden handle from Stenild, Denmark (after Steensberg). (3) Hypothetical hafting of crescentic flint sickle (after Steensberg). Not to scale.

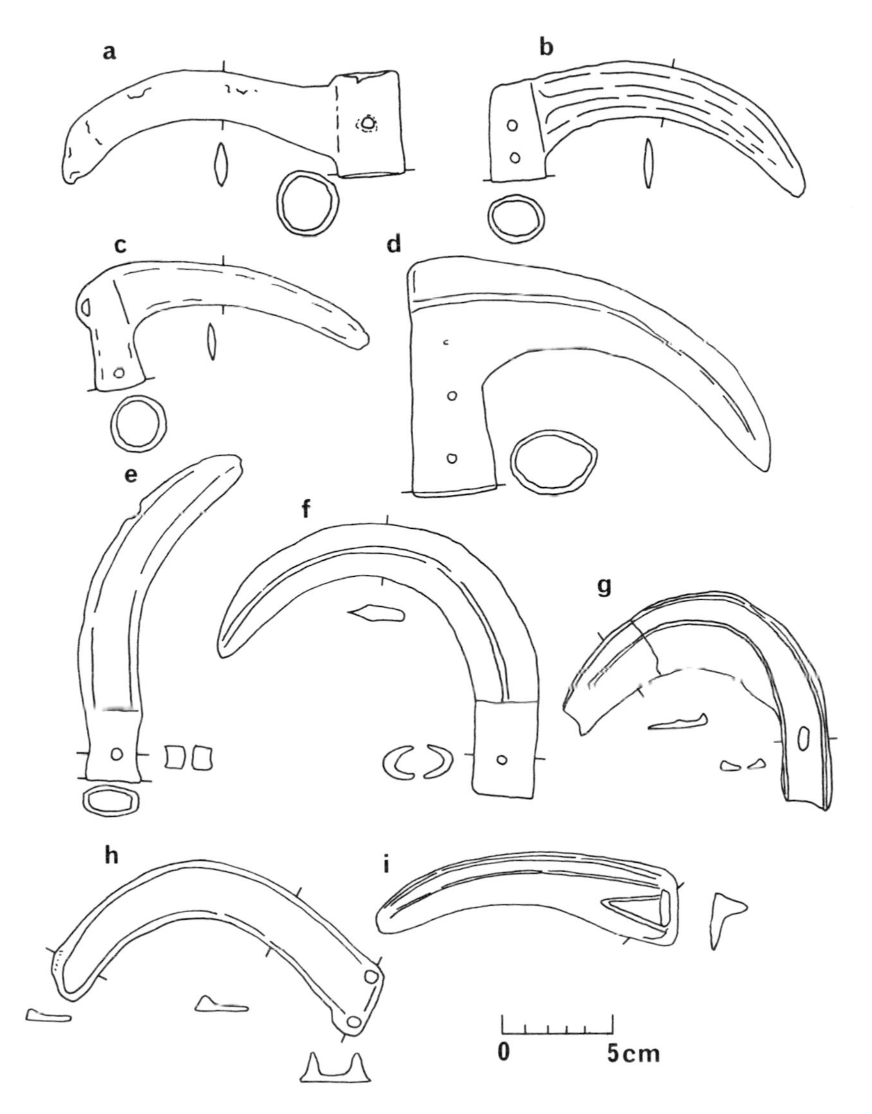

Fig. 23. (a-f) Socketed bronze sickles from Downham Fen, Streatham Fen, Winterbourne Monkton, Llyn Fawr, Rosebury Topping and London. **(g)** Riveted sickle from Thames at Taplow. **(h, i)** Knobbed sickles from Edington Burtle.

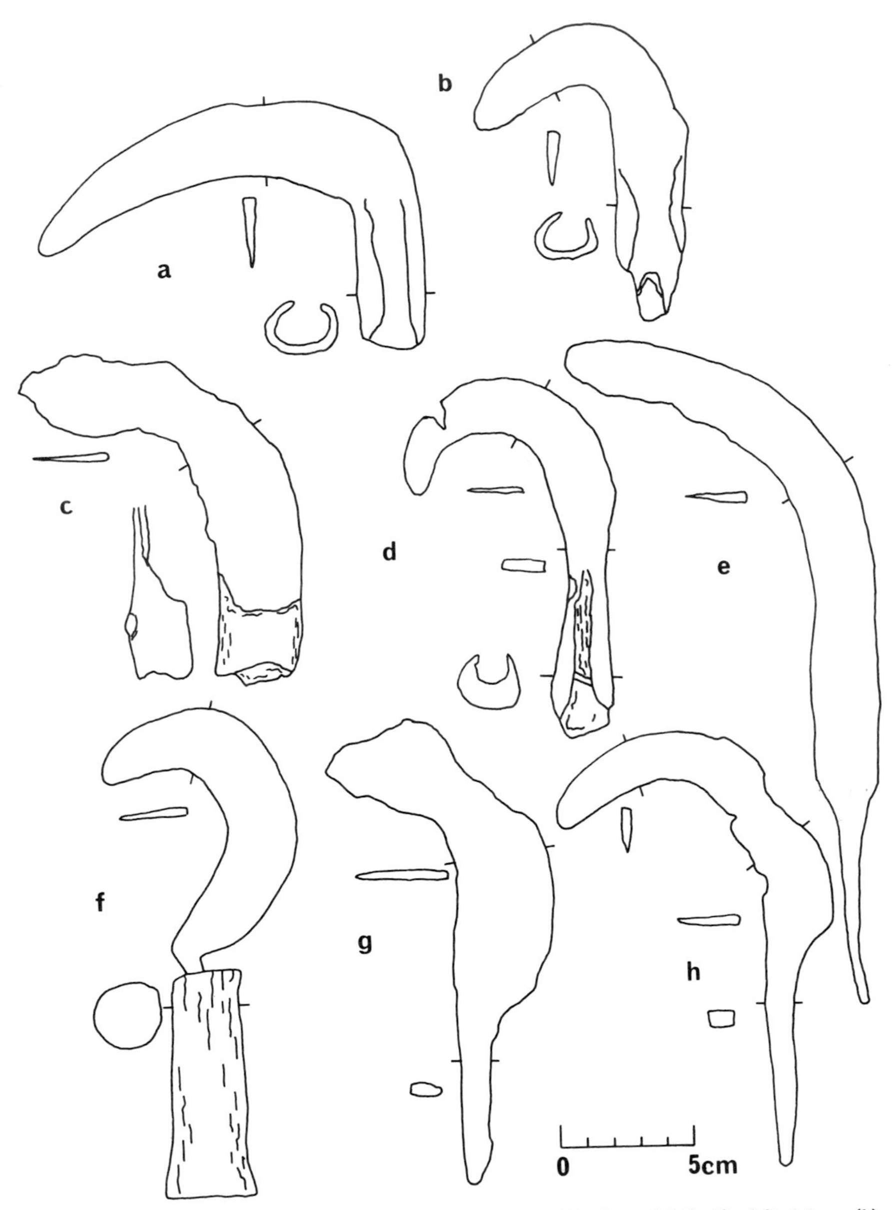

Fig. 24. Iron-age and Romano-British iron hooks and sickles from **(a)** Codford St. Mary, **(b)** Hambledon Hill, **(c)** Maiden Castle, **(d)** Caerwent, **(e)** Wilderspool, **(f)** Linton, **(g)** Coygan Camp and **(h)** Ham Hill.

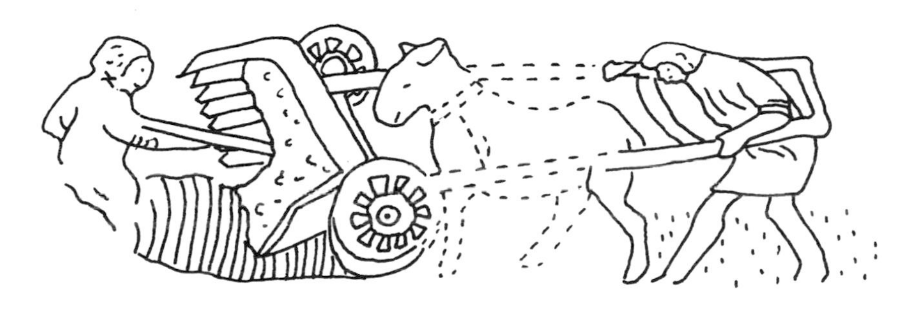

Fig. 25. Two forms of the Gallo-Roman harvesting machine. **(a)** The Palladian machine as reconstructed by Quimper. **(b)** The machine as described by Pliny, as shown by the fragments of relief sculpture from Buzenol and Arlon. (Both after White.)

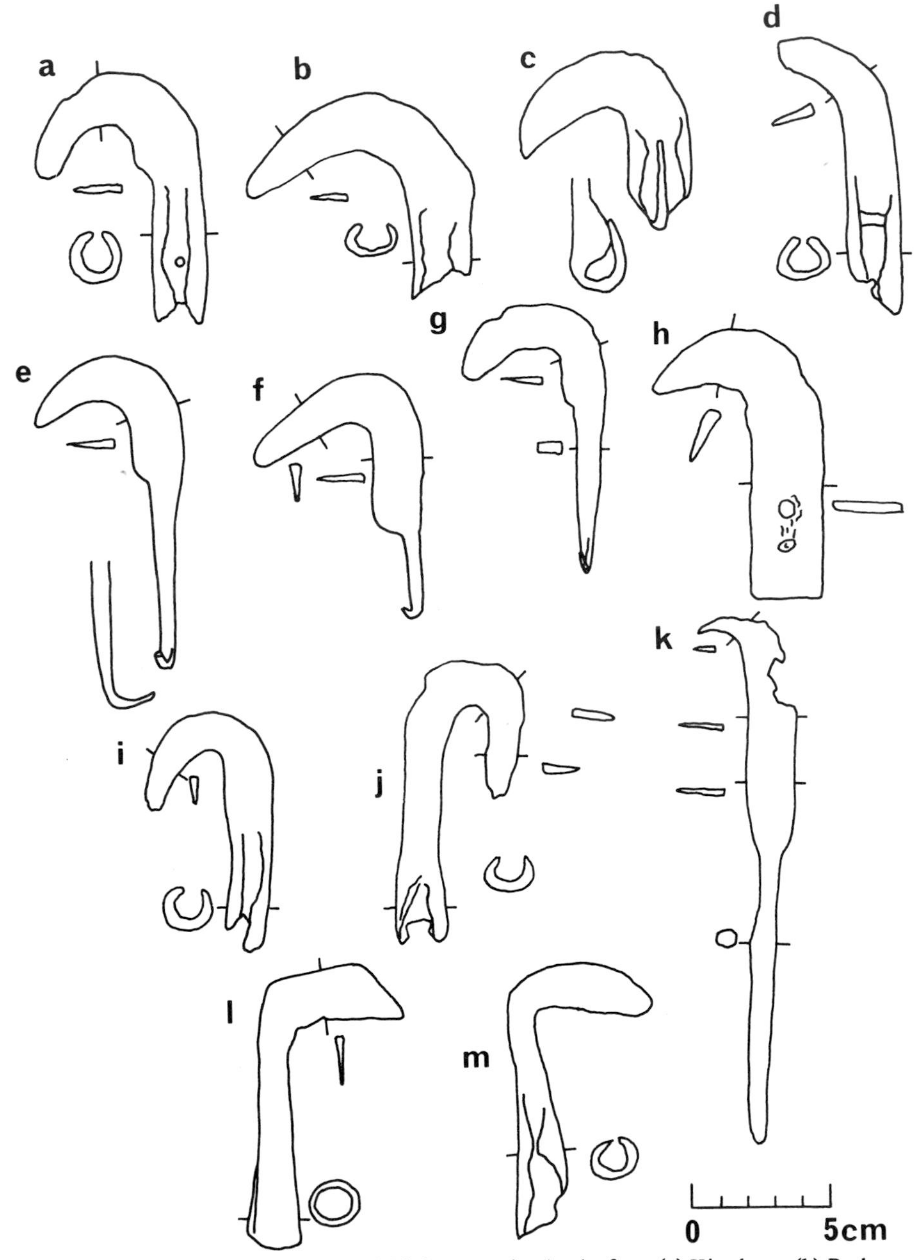

Fig. 26. Iron age and Romano-British iron pruning hooks from **(a)** Kimsbury, **(b)** Barbury Castle, **(c)** Fifield Bavant, **(d)** Kelvedon, **(e, f, g)** Caerwent, **(h)** Aldwick, **(i)** Salmonsbury, **(j)** Kettering (?), **(k)** Caerwent, **(l)** Rushall Down, **(m)** Silchester.

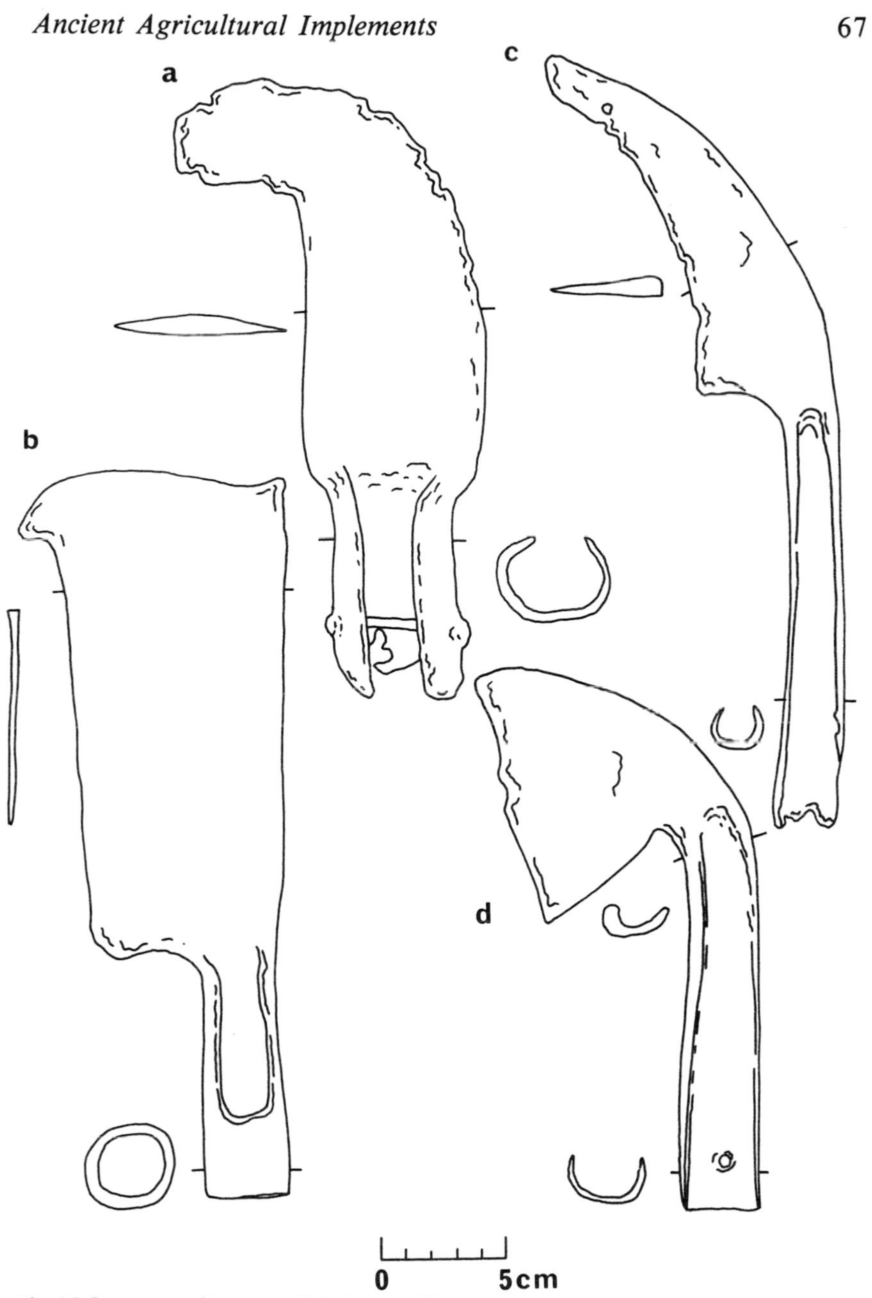

Fig. 27. Iron age and Romano-British iron billhooks from **(a)** Glastonbury, **(b)** Colchester, **(c)** Newstead, **(d)** Cirencester.

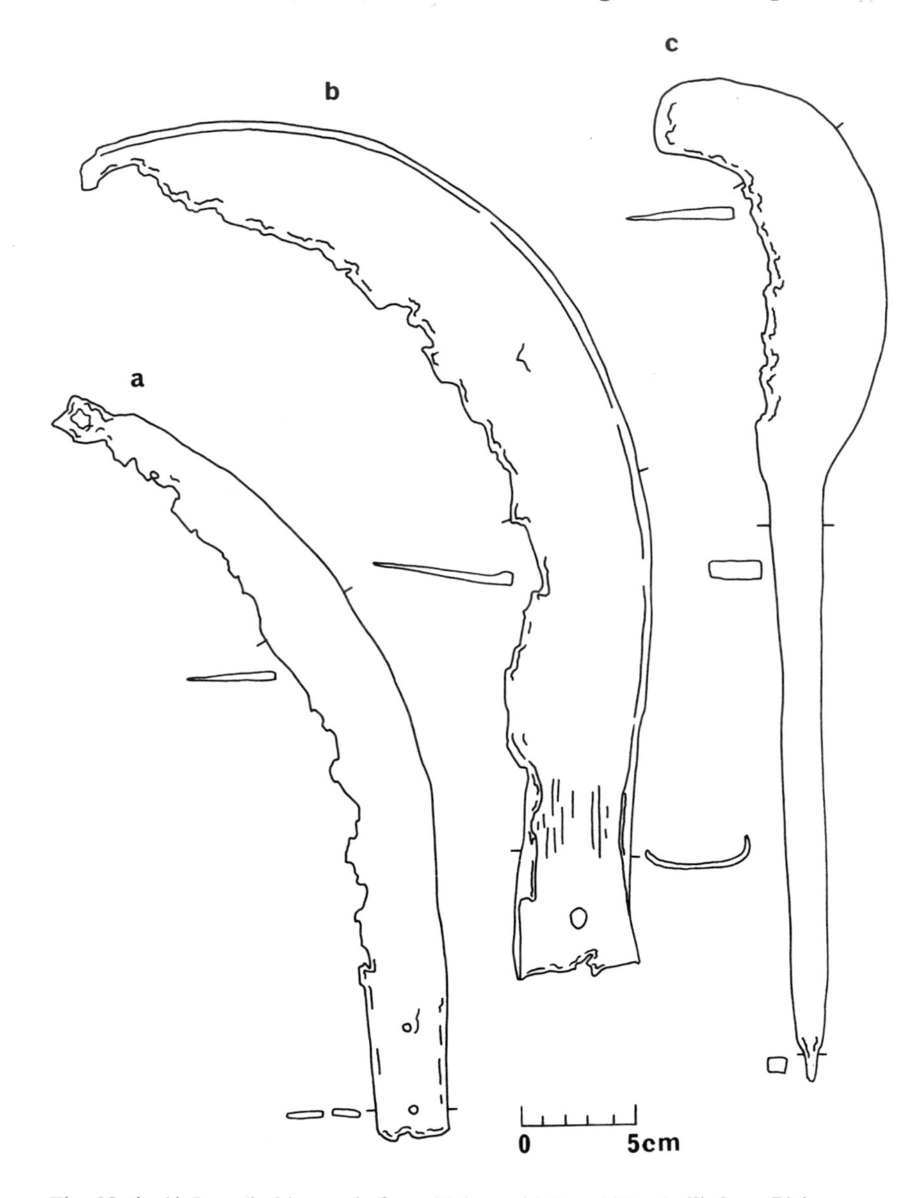

Fig. 28. (a, b) Iron slashing tools from Bigbury. **(c)** Iron billhook (?) from Bigbury.

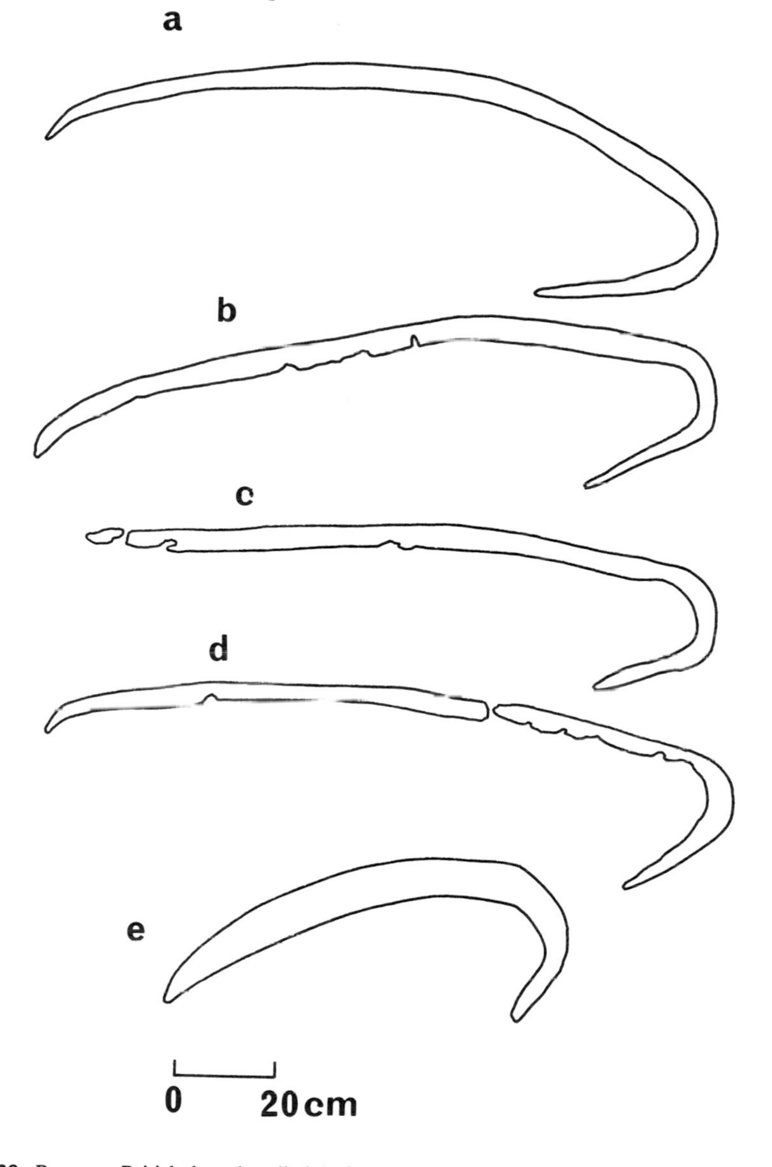

Fig. 29. Romano-British long-handled balanced scythes from **(a)** Great Chesterford, **(b)** Abington Pigotts, **(c)** Farmoor, **(d)** Hardwick, **(e)** Newstead.

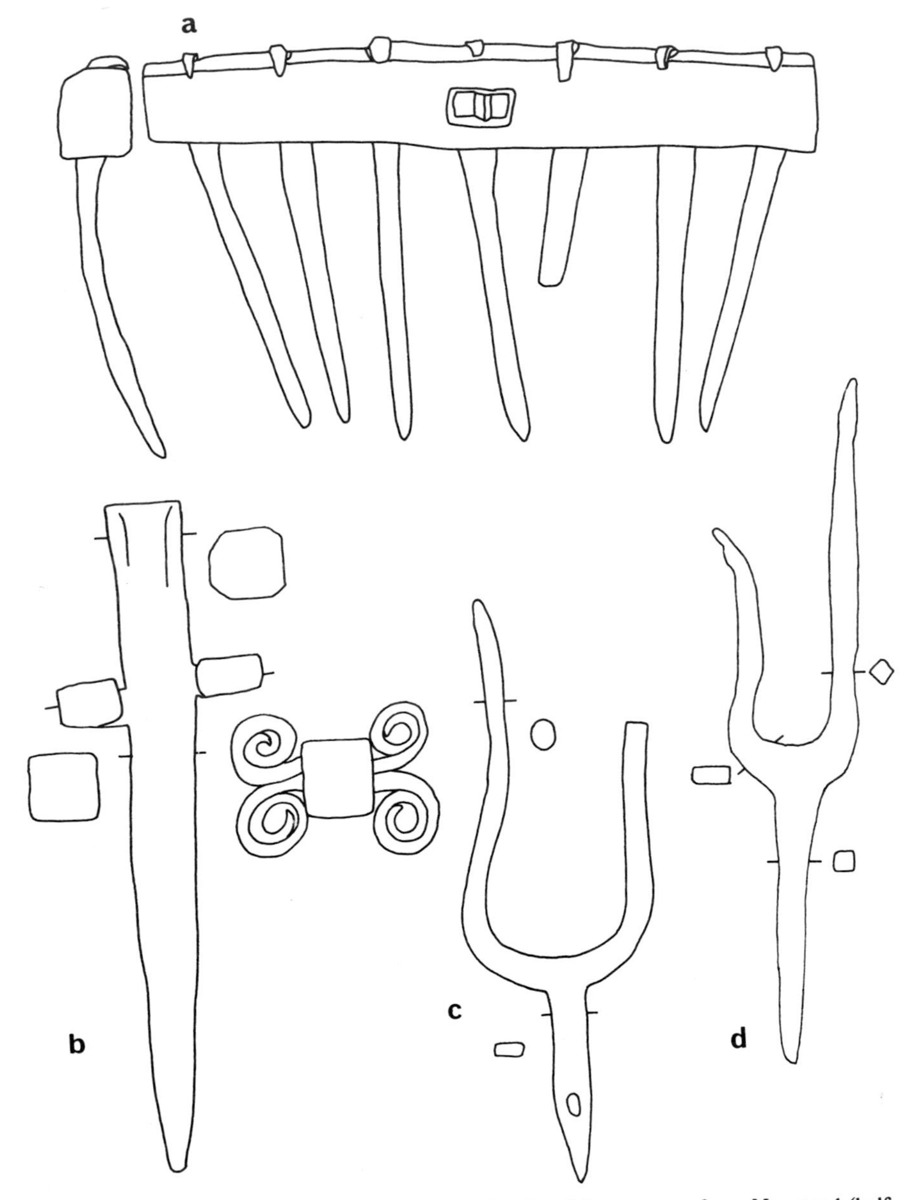

Fig. 30. (a) Romano-British rake with wooden head and iron prongs from Newstead (half size). **(b)** Romano-British iron mowers' anvil from Silchester (3/5 size). **(c, d)** Romano-British iron pitchforks from Newstead and Silchester (3/5 size).

Index